Nuclear Energy and Liability in South Asia

M.P. Ram Mohan

Nuclear Energy and Liability in South Asia

Institutions, Legal Frameworks and Risk Assessment Within SAARC

Springer

M.P. Ram Mohan
TERI University
New Delhi
India

ISBN 978-81-322-2342-9 ISBN 978-81-322-2343-6 (eBook)
DOI 10.1007/978-81-322-2343-6

Library of Congress Control Number: 2015934950

Springer New Delhi Heidelberg New York Dordrecht London

Printed on acid-free paper

Springer (India) Pvt. Ltd. is part of Springer Science+Business Media
(www.springer.com)

Preface

In the development of international law relating to liability, the nuclear energy sector represents an alternative approach to transboundary liability regime. Building on this foundation and following the Chernobyl accident—international consensus was sought for a stronger transboundary legal regime in the event of a nuclear disaster. However, after 60 years of the existence of international nuclear liability laws and 25 years after Chernobyl, the primary objective of the Conventions—harmonisation and a global regime remains unfulfilled. Further, many countries are expanding or introducing nuclear programmes, without adequate transboundary legal protection. On account of these issues, a regional approach to nuclear risk framework appears more promising than the non-achievable global regime. South Asia, with its rapidly expanding nuclear energy footprint is in a unique position to adopt a regional mechanism.

This study undertaken during the period 2009 to 2014, brings out the difficulties that lie ahead in achieving an overarching global nuclear liability architecture. In the case of South Asia—a densely populated region, the existing legal regime and legal arrangements which are under consideration are unlikely to secure a transboundary liability remedy. The technical risk assessment study points to the likely transboundary impact in case of a nuclear accident. There has not been any attempt in forging a regional consensus on the issue of nuclear energy risk to date. The book argues that a regional approach is certainly possible in South Asia, under the structure of SAARC, and the expert opinions confirm this proposition. The result gives a basic framework for decision-makers in SAARC on implementing measures that address the transboundary nuclear energy risk concern.

Keywords: Nuclear liability, Compensation, Transboundary pollution, International legal regime, South Asia, SAARC, Nuclear risk community, Risk mapping, Risk perception

Acknowledgments

My interest in 'nuclear energy and law' arose purely by chance. As a research assistant in 2000 I was asked by Prof. N.R. Madhava Menon, then Vice-Chancellor, National University of Juridical Sciences, Kolkata to prepare a syllabus on energy law. I am indebted to Prof. Menon for his encouragement over the years, and in particular his support for the nuclear law training programme at the University of Montpellier, Paris organised by OECD-Nuclear Energy Agency. The course opened a world of opportunities, and most importantly—was introduced to Patrick Reyners. Over the years Patrick guided my forays into this field of law.

Even though nuclear energy has never been a 'topic' for 'critical' discussion in many institutions in India, my employer—TERI, provided exceptional backing for many of my initial writings and research proposals on the subject. Dr. R.K. Pachuari, head of TERI has institutionally supported my doctoral studies; Dr. Leena Srivastava, allowed me to experiment, and provided an invisible hand of support to many of my activities, which continues even today. I am grateful to TERI, my employer of 10 years in having a liberal study leave policy.

I was most fortunate in my interaction with two of India's foremost nuclear scientists—Dr. R. Chidambaram and Dr. R.B. Grover. They were always welcoming; I am truly thankful to them.

Undertaking doctoral studies at the Indian Institute of Technology, Kharagpur (IIT-KGP), an institution of immense history and importance to India, was a privilege. The Rajiv Gandhi School of Intellectual Property Law at IIT-KGP was established to study the interaction between technology and laws; and my Ph.D. topic, 'Nuclear Energy and Liability' was a good fit. My supervisor, Dr. K.D. Raju provided all assistance in making me feel at home. Apart from an international law expert, Dr. Raju was a military man too. That meant even though I was given full academic freedom, he ensured deadlines were equally strict. M.V. Shiju a friend of many years, whole heartedly supported my endeavour. I was unsure about joining the programme at the time, he was kind enough to give me his train ticket to Kolkata which was purchased to celebrate his wedding anniversary. I am proud to have a friend of his stature and humbleness.

The support by Dr. P.K. Joshi, Dr. G. Mini, Anusheema Chakaborty and Gopal Sarangi from TERI University and TERI has been invaluable. They spent long hours helping me, removing my doubts, helping plot risk zones, creating maps and devising research methodology. The work was a result of numerous interviews I have undertaken, without naming anyone in particular I am profusely thankful to everyone who generously shared their thoughts.

One of the most visible outcomes of the time spent on study leave is the establishment of Nuclear Law Association, India (NLA). The extremely competent first Governing Board of NLA consisting of Arun Khanna, Els Reynaers Kini, Mohit Abraham, Dr. Rajesh Babu, Dipankar Bandyopadhyay and T.P. Krishna believed in me in creating a unique association and committed enormous time and resources.

I was able to do my Ph.D. only because Praachi, my wife, encouraged and supported even when there were small children to take care at home. Many months away from home, long hours of work and my erratic behaviour while writing the thesis, must have taken a toll on her. She can fully justify that the thesis is equally hers, considering the time she spent on reviewing, correcting and debating issues. I am glad that I could finish the work before Uday and Utsa, could start demanding my presence. Amma, if she was alive would have been really proud and Achan was the strongest supporter in my life. Finally, I am indebted to Ma and Papa, and brothers—Mahesh, Paramesh and Prateek for unstinted love.

Contents

About the Author

Dr. M.P. Ram Mohan is Associate Professor at the Department of Policy Studies, TERI University and Adjunct Fellow at The Energy and Resources Institute (TERI), New Delhi, India. Dr. Mohan is the founding President of the Nuclear Law Association, India and is part of many of the governmental and non-governmental deliberations on nuclear energy programme, law and policies. He is a nominated member of the "Nuclear energy for peaceful applications committee", Bureau of Indian Standards, Government of India. Ram Mohan's areas of work include infrastructure and regulatory law, nuclear technology Law, energy and environmental law. IUCN has recognised him as the India Focal point of IUCN World Commission on Environmental Law and he is a member of its Energy Law and Climate Change Specialist Group. Prior to TERI, he worked at the National University of Juridical Sciences (NUJS) as Research Faculty and Coordinator, School of Technology, Law and Development. He started his career as an advocate practising at the Trivandrum District Court, Kerala. He is a yearly contributor to Yearbook of International Environmental Law published by OUP; Managing Editor, Journal of Law and Global South; and is on the editorial board of International Journal of Nuclear Law (IJNucL) and Atoms for Peace: An International Journal (AFP). He is also the India communicator to OECD—Nuclear Law Bulletin.

Symbols and Abbreviations

AEA	Atomic Energy Act, 1962 (India)
AEC	Atomic Energy Commission
AERB	Atomic Energy Regulatory Board
BAEC	Bangladesh Atomic Energy Commission
CANDU	Canadian Deuterium Reactor
CEIP	Carnegie Endowment for International Peace
DAE	Department of Atomic Energy
EIA	Environmental Impact Assessment
EU	European Union
FBR	Fast Breeder Reactors
FNCA	Forum for Nuclear Co-operation in Asia
GIS	Geographical Information Systems
HBNI	Homi Bhabha National Institute
IAEA	International Atomic Energy Agency
ICJ	International Court of Justice
KANUPP	Karachi Nuclear Power Plant
km	Kilometer
KNPP	Kudankulam Nuclear Power Plant
MoEF	Ministry of Environment and Forest (India)
MW	Megawatt
NAPS	Narora Atomic Power Station
NDMA	National Disaster Management Authority
NNPAP	National Nuclear Power Action Plan (Bangladesh)
NNWS	Non-Nuclear Weapons States
NPCIL	Nuclear Power Corporation of India Limited
NPP	Nuclear Power Plant
NSG	Nuclear Suppliers Group
NWS	Nuclear Weapons States
OECD	Organization for Economic Co-operation and Development
OECD-NEA	OECD Nuclear Energy Agency
OEEC	Organization for European Economic Co-operation

PAEC	Pakistan Atomic Energy Commission
PAZ	Precautionary Action Zone
PHWR	Pressurised Heavy Water Reactor
SAARC	South Asian Association for Regional Cooperation
SDMC	SAARC Disaster Management Centre
SDR	Special Drawing Rights
UK	United Kingdom
UN	United Nations
UNAEC	United Nations Atomic Energy Commission
UNSCEAR	United Nations Scientific Committee on the Effects of Atomic Radiation
UPZ	Urgent Protective Action Planning Zone
US	United States of America
US AEA	United States Atomic Energy Act
USAEC	United States Atomic Energy Commission
USDOE	United States Department of Energy
USNRC	United States Nuclear Regulatory Commission

List of Statutes and International Conventions

1. The Bangladesh Atomic Energy Regulatory Act, 2012.
2. The Convention on Assistance in the Case of a Nuclear Accident or Radiological Emergency, 1986.
3. The Convention on Early Notification of a Nuclear Accident, 1986.
4. The Convention on Supplementary Compensation for Nuclear Damage, 1997.
5. The Convention Supplementary to the Paris Convention of 29 July 1960.
6. The Joint Protocol Relating to the Application of the Vienna Convention and the Paris Convention, 1988.
7. The Nuclear Installations Act, 1965 (United Kingdom).
8. The Nuclear Liability Act, 2012 as amended (Canada).
9. The Nuclear Non-proliferation Act of 1978 (United States).
10. The Nuclear Non-Proliferation Treaty (NPT), 1968.
11. The Paris Convention on Third Party Liability in the Field of Nuclear Energy, 1960.
12. The Price-Anderson Nuclear Industries Indemnity Act, 1957.
13. The Protocol to Amend the 1963 Vienna Convention on Civil Liability for Nuclear Damage, 1997.
14. The Protocol to Amend the Brussels Convention Supplementary to the Paris Convention, 2004.
15. The Protocol to Amend the Paris Convention on Third Party Liability in the Field of Nuclear Energy, 2004.
16. The Atomic Energy Act, 1 August, 1946.
17. The Atomic Energy Act, 1962.
18. The Civil Liability for Nuclear Damage Act, 2012.
19. The Environmental Protection Act, 1986.
20. The Nuclear Non-proliferation Act, 1978.
21. The Public Liability Insurance Act, 1991.
22. The Third Party Liability for Nuclear Damages, 1998 (Austria).
23. The Vienna Convention on Civil Liability for Nuclear Damage, 1963.

List of Cases

1. *Carolina Environmental Study Group v. Atomic Energy Commission (438 U.S. 59 (1978))*
2. *Case Concerning Military and Paramilitary Activities in and Against Nicaragua (Nicaragua v. United States of America) (1986 I.C.J. 14)*
3. *Centre for Public Interest Litigation and others v. Union of India (WP 407/2012)*
4. *Common Cause and others v. Union of India (WP 464/2011)*
5. *Conservation Society of Southern Vermont v. Atomic Energy Commission (No. 19-72 (D.D.C. April 16, 1975))*
6. *G. Sundarrajan v. Union of India (W.P.Nos. 24770 and 22771 of 2011, 8262 and 13987 of 2012 and W.P.(MD) Nos. 14054 and 14172 of 2011, 1823 and 2485 of 2012)*
7. *Indian Council of Enviro-Legal Action v. Union of India (AIR 1996 SC 1466)*
8. *M.C. Mehta v. Union of India (Oleum Gas Leak Case) AIR 1987 SC 1086)*
9. *Ryland v. Fletcher (L.R. 3 H.L. 330; [1861–1873] All E.R)*
10. *Case concerning the Aerial Incident of 10 August 1999 (Pakistan v. India) (ICJ. Reports 2000)*
11. *Trail Smelter Arbitration (United States Vs Canada) 1941, U.N. Rep. International Arbitration Awards 1905 (1949)*
12. *Corfu Channel Case (1949). (I.C.J. Reports 1949)*

Chapter 1
Introduction

Abstract This study brings out the difficulties that lie ahead in achieving global nuclear liability architecture. In the case of South Asia—a densely populated region, the existing legal regime and legal arrangements which are under consideration is unlikely to secure a transboundary liability remedy. The technical risk assessment study points to the likely transboundary impact in case of a nuclear accident. There has not been any attempt in forging a regional consensus on the issue of nuclear energy risk to date. The book argues that a regional approach is certainly possible in South Asia, under the structure of SAARC, and the expert opinion confirms this proposition. The result gives a basic framework for decision-makers in SAARC on implementing measures that addressing the nuclear energy risk concern.

Keywords Nuclear liability literature · Review of nuclear liability law development · Nuclear energy and South Asia · Critical analysis of transboundary nuclear liability regime

1.1 Background

The adoption of nuclear energy for power generation by countries worldwide has been a contentious issue. Contentious for the fact that, development of nuclear energy was initiated for military purposes and the dreadful consequences of the bombing of Hiroshima and Nagasaki, Japan, on 6 and 9 August 1945 has contributed to the continued perception of fear. However, countries have moved forward in their adoption of this technology, mainly due to large possibilities in generating abundant power. But the question remains whether this forward march is without understanding the larger implications of nuclear energy technology adoption.

M.P. Ram Mohan, *Nuclear Energy and Liability in South Asia*,
DOI 10.1007/978-81-322-2343-6_1

These concerns were not unfounded either. Three-Mile Island nuclear accident in the United States (US) (1979),[1] Chernobyl in the former Soviet Union (USSR) (1986)[2] and the nuclear fallout subsequent to earthquake and tsunami in Fukushima, Japan (2011)[3] exposed the nature of nuclear accidents and our helplessness to grapple with the accidents of such magnitude.

For the above disasters, in terms of transboundary impacts—the subject of the book, the Chernobyl accident as reported resulted in considerable human, environmental and economic loss, both in the country of origin and much of Europe. More importantly, in the Chernobyl case, the offending State failed to immediately protect, assist or otherwise notify any of its neighbouring States.

The Chernobyl accident clearly illustrated that the geographical scope of nuclear damage is not necessarily confined to national boundaries. Official acknowledgement of the accident from Soviet Union came only much later. It was the Swedish authorities who raised the alarm and notified the international community on the occurrence of the disaster. Soviet Union, at that time was not a party to any of the international conventions relating to nuclear liability; because of this there were no legal requirements for payment of international compensation.

This experience led the international community to believe that an international consensus on 'state liability' in the event of a nuclear accident is not only desirable but necessary. In response to Chernobyl, the world community over the past few decades, has endeavoured through international legal instruments to impose stricter obligations upon States pursuing civil nuclear energy programmes.

However, 25 years after the Chernobyl accident, even today, it appears that the international consensus has not been achieved on many issues, and a unified international legal regime—the foundation of successful nuclear energy programmes, remains elusive. Most of the existing international conventions are not adequately adhered to, and national laws are at variance with each other and also with the international conventions. The hard fact is that the countries with a majority of the operating nuclear power plants (including Canada, China, Japan, Korea, South Africa, Switzerland and the US) have only recently considering or are yet to be part of an international regime. Instead, many rely on their own domestic liability laws.

---

[1] On March 28, 1979 a partial nuclear meltdown occurred at the Three Mile Island power plant in Dauphin County, Pennsylvania, United States. It was the worst accident in US commercial nuclear power plant history, and resulted in the release of small amounts of radioactive gases and radioactive iodine into the environment. There was no transboundary consequence from the accident.

[2] On April 26 1986, a testing went horribly wrong, causing an explosion at the Chernobyl nuclear power station in northern Ukraine. The Chernobyl disaster is the worst nuclear power plant accident in the history, and the first one to be classified as a level 7 event (extreme) on the IAEA International Nuclear Event Scale. The impact was considerable not only to the people and environment in and around Ukraine, but to whole of Europe.

[3] The 2011 nuclear accident in Fukushima Japan is the result of series of natural events—earthquake and then tsunami leading to nuclear accident. IAEA International Nuclear Event Scale categorises it as level 7.

The demand for alternative power sources has led to the revival of nuclear energy programmes in the last decade. These new programmes have been initiated on the assurance that the nuclear industry today is technically far superior; and importantly for the purposes of this study, supported by well-defined nuclear liability and compensation laws. It is articulated that the legal regime as it stands today, is in a position to respond to the challenges of Chernobyl type accidents.

There is no doubt that there have been significant legal developments on the question of liability, but this study argues that the overall legal and institutional response is far from satisfactory. This question is of great significance on account of the recent large-scale development of nuclear power plants, particularly in the South Asian region. The danger is that developing countries are embracing nuclear energy, with little or no legal safeguards to fall back to, in the case of a disaster. But to be fair, undoubtedly, given the complexities, efforts of international institutions and legal community have been noteworthy.

1.2 International Law and Nuclear Liability

Under the auspices of Organisation for Economic Cooperation and Development (OECD), *The Convention on Third Party Liability in the Field of Nuclear Energy* 1960 (Paris Convention), established the nuclear liability regime for most of Western Europe. Supplementing the Paris Convention, *the 1963 Convention Supplementary to the Paris Convention of 29 July 1960* (Brussels Supplementary Convention) was established to provide for enhanced compensation. The International Atomic Energy Agency (IAEA) sponsored an international regime in 1963—the *Vienna Convention on Civil Liability for Nuclear Damage* (Vienna Convention), largely on the principles laid down in the Paris Convention.

The principles laid down by the Paris and Vienna Conventions now form the fundamental principles of international nuclear liability law. The principles detailed below, remain the bed rock of State responsibility and liability. Many countries have embraced these principles in their own domestic legislations as well. The broad principles are:

First, the Conventions have a no-fault liability (absolute liability);

Second, the Conventions channel liability exclusively to the operator of the nuclear installation (legal channeling);

Third, the Conventions mandate only courts of the State in which the nuclear accident occurs have jurisdiction (exclusive jurisdiction);

Fourth, the Conventions provide limitation of the amount of liability and the time for claiming damages (limited liability);

Fifth, the Conventions provide limitation of time for claiming damages (limitation in time);

Sixth, the operator must secure insurance or financial guarantee to the extent of his liability amount.

The effectiveness of these Conventions has been continuously doubted because of: (1) the non-effectiveness of these regimes, as many large nuclear energy producing countries remain outside the Paris and Vienna Conventions; (2) drafting of national laws substantially differently from the provisions of the Conventions, thus hitting a road block towards the efforts in harmonisation; (3) varying industry structure in several countries, thus making legal channeling a contentious issue; and (4) the difficulty in bringing countries with unlimited and limited liability requirements together as far as compensation regime is concerned.

The Chernobyl accident in a significant way, and adoption of nuclear energy by many countries in the past few decades, tested these instruments in all practicality and found them grossly inefficient.

Subsequent to Chernobyl, the European Union (EU) undertook a complete revision of Paris and Brussels Conventions. The "*2004 Protocol to Amend the Paris Convention on Third Party Liability in the Field of Nuclear Energy of 29 July 1960*" (2004 Protocol) that amended both the Paris and Brussels Conventions is the most important of these revisions. The protocol extends the geographic scope of accident, raises the amount of compensation available to the victims and enlarges the definition of the nuclear damage. The protocol recognised for the first time the importance of coexistence of limited and unlimited liability regimes. However, the 2004 Protocol has not yet entered into force, since a majority of the EU states including Germany, France, and the United Kingdom have not ratified it, thereby permitting the pre-Chernobyl legal situation to continue.

The Vienna Convention has also undergone significant changes. In 1997, over 80 States adopted a Protocol to amend the 1963 Vienna Convention on Civil Liability for Nuclear Damage (1997 Protocol) which came into force in 2003. Overall, the Vienna Convention, which was aimed at universal adherence, could only attract memberships of 36 states till date. With both these Conventions being independent of each other, it was open to the States to adopt either of these conventions. This raised the issue of coordination and harmonisation between the above two Conventions. To rectify this defect, in 1988 the two main conventions were linked by the *Joint Protocol Relating to the Application of the Vienna Convention and the Paris Convention* (1988 Joint Protocol) which came into force in 1992. However, interestingly many States in Europe—a region affected by Chernobyl have not ratified the Joint Protocol, including Belgium, Switzerland, the UK, etc.—major nuclear power producing countries in Europe. France ratified the Joint Protocol only in 2014. An example from Europe would give a better picture. In 2005 Russia ratified the Vienna Convention. Russia being a non-member of OECD did not adopt the Paris Convention. Russia also did not adopt the 1988 Joint Protocol. If a situation similar to Chernobyl arises, Russia could still argue that they are not a party to the Paris Convention and Joint Protocol, and avoid compensating neighbouring Paris Convention states.

In addition to the Paris and Vienna Conventions, in 1997 at the instance of US, IAEA sponsored another international nuclear liability regime—Convention on Supplementary Compensation (CSC or Compensation Convention). CSC was largely brought in place to align the US domestic legislation-Price-Anderson

Nuclear Industries Indemnity Act (1957) (The Price Anderson Act) with international law. It is structured as an umbrella legislation. CSC provides additional amounts to be offered through contributions by the State parties on the basis of installed nuclear capacity. As of now, CSC has six Contracting States and 19 Signatories. The US is a party to the Convention. With Japan's ratification on 15 January 2015, the Convention is effective from 15 April 2015. The advantage of CSC is that this is an instrument to which all States may adhere to, regardless of whether they are parties to any existing nuclear liability conventions or not, or whether they have nuclear installations on their territories or not. However, an important requirement is that the countries should enact/amend national liability laws consistent with the annex (model law) to the CSC.

An overview of nuclear power programmes in South Asia reveals that the SAARC may also face similar issues that were witnessed in other nuclear power regions of the world.

1.3 South Asia

IAEA (2007a, b) estimates that on account of the massive energy requirements to fuel the economic progress and to meet the demands of a large population, Asia may well be the engine of the world's nuclear energy growth. Even though IAEA (2011a–c) revised its projection estimates subsequent to the Fukushima accident in Japan, the major portion of global expansion of nuclear power is still projected to be in Asia. South Asia (Afghanistan, Bangladesh, Bhutan, India, Pakistan, Nepal, Maldives and Sri Lanka)—the subject of this thesis, is a unique region in terms of nuclear legal architecture—the Nuclear Non-Proliferation Treaty (NPT), 1970. Except Israel, India, Pakistan and North Korea all other countries of the world are party to NPT. Of these India and Pakistan are in South Asia; both have energy as well as weapon programmes. Currently, India and Pakistan are expanding their share of nuclear energy, and Bangladesh and Sri Lanka have plans to start a nuclear energy programme. Many of the proposed Nuclear Power Plants (NPPs) are geographically not far from each other. Considering the transboundary impacts of Chernobyl, and more recently in Fukushima, these projects in South Asia could be seen as a shared concern in the region.

In respect of international legal participation, except India, none of the SAARC countries are party to the Vienna or any other international liability convention that are in force, thereby exposing the region to a probable "Chernobyl like" situation. What happened subsequent to the Chernobyl accident, i.e. avoiding State liability, could happen with these South Asian countries as well. A country could easily reject liability and refuse payment of any compensation. India became party to CSC in 2010. Bangladesh being a member of NPT will become a part of the Vienna convention. In terms of domestic legislation or international ratification, only Bangladesh (*Bangladesh Atomic Energy Regulatory Act, 2012*) (Bangladesh liability law) and India (*The Civil Liability for Nuclear Damage Act, 2010*)

(Indian liability law) have domestic nuclear liability laws. Even if we take the example of India and Bangladesh, these countries' adherence to different conventions may lead to future harmonisation issues in respect to jurisdiction.

Nevertheless, all SAARC countries are parties to *The Convention on Assistance in Case of a Nuclear Accident or Radiological Emergency, 1986* and *the Convention on Early Notification of a Nuclear Accident, 1986* (Emergency Conventions) albeit with some reservations. However, these conventions do not dwell into any legal obligations for parties to accept its responsibility for liability and compensation.

Interestingly though, there are already some discussions and concerns on the transboundary consequences of nuclear energy developments. Sri Lanka in 2012 reportedly raised and subsequently retracted its transboundary concern emanating from India's Kudankulam Nuclear Power Plant (KNPP), with IAEA. From the initial stages of construction of KNPP itself, Sri Lanka had apprehensions of safety, since the country is only 250 km away from the Tamil Nadu coast. The nuclear power project in Bangladesh which is only 50 km from the Indian border could be a future contentious issue. There are several instances of risk, if one considers the existing, proposed and planned projects in South Asia.

South Asia in short, will be a test case as to how legal institutions and regional mechanisms develop, and respond to the challenges of nuclear energy programmes. The adherence to nuclear liability regime as it exists today, does not seem to give hope for either strengthening the international liability regime or having a South Asian regional consensus.

1.4 National Laws on Nuclear Liability

Beyond the international and regional conventions, several countries with commercial nuclear programmes have their own legislative regimes for nuclear liability. There are three categories of countries in this regard: (1) those that are party to one or both of the international conventions and have their own legislation (2) those that are not parties to any international convention that is in force but have their own legislation and (3) those that are not party to any convention and are without their own legislation.

In Europe, most countries have enacted laws that are largely consistent with the Paris/Brussels Conventions, but with varying schemes of liability in terms of liability amount and recourse. However, what is being witnessed in Europe is that, mostly West European countries follow the Paris Convention and most of the East European countries follow the Vienna Convention. The reasons for such clear distinction within Europe are historical.[4]

[4] The Organisation for European Economic Co-operation; (OEEC) established on 16 April 1948 was with the purpose of post-war recovery programme. OEEC is now renamed as OECD.

The US has its own unique liability regime—The Price-Anderson Nuclear Industries Indemnity Act (1957) (the Price-Anderson Act 1957) and the CSC are structured in such a way as to protect the Price Anderson Act through grandfather clause treatment. The Act adopts what is called economic channeling[5] of liability unlike legal channeling followed elsewhere. In terms of transboundary impacts, US law provides for reciprocity. This is the same as the *Canadian Nuclear Liability Act, 2012,* as amended (Canadian liability law). Thus, currently between US and Canada, with regard to a nuclear accident having transboundary impact, it would be the respective national laws that would determine issues relating to liability, channeling, etc. Canada is also expected to join CSC soon.

In South Asia, India is committed to aligning itself with the CSC. However, while enacting the Indian liability law, it made few deviations from the CSC mandate which are discussed below.[6] Bangladesh, the other country in South Asia having a liability regime, follows the Vienna Convention. Indian law has been criticised on grounds of inconsistency with CSC and thus opens more debate on nuclear liability laws.[7]

The study assumes significance in the wake of a new wave of nuclear resurgence sweeping around the world, largely led by the high-population countries of Asia. Also, several European countries which are committed to nuclear phase-outs, are reconsidering it on account of both climate change concerns and fortifying their economic growth from future non-availability of conventional fuels. The example of the Indian liability law provides an opportunity to reassess the liability regime in the context of emerging countries in respect of national interest and transboundary concerns. In the aftermath of the Fukushima, IAEA's (2011a–c, 2012) call for overhaul and rationalisation of the several international conventions provides impetus to this study.

The international nuclear liability regime today remains complicated and features sparse participation. National laws, on the other hand, are different from each other and evade major questions on transboundary liability. The study focuses on the nuclear liability and compensation laws developed internationally and regionally, and discuss the operative arrangements of these instruments.

---

[5] "Legal channeling means that all liability is channelled to the nuclear operator; no other entity may be held liable for nuclear damage. Economic channeling means that any entity may be held legally liable, but the economic consequences of that liability are channelled to the liable nuclear operator. Thus, any person who is held legally liable will be indemnified in respect of that liability" (Schwartz 2006).

[6] Deviations were particularly in two aspects. The first *one* is on the limitation to liability and the second related to the right to recourse. Though there has been some discussions on India's divergent stance on CSC, provision relating to an option of seeking further remedies was retained. Section 46 of the Indian law provides that *"the provisions in the Act shall be in addition to, and not in derogation of, any other law for the time being in force, and nothing contained herein shall exempt the operator from any proceeding which might, apart from this Act, be instituted against such operator".*

[7] Indian law is argued as inconsistent mainly on, liability amounts and claims, and right of recourse. These are discussed in detail in Chap. 3.

1.5 Overview of the Critical Literatures on Transboundary Nuclear Liability Regime

Many of the critical writings on liability law are on the environmental and human harm arising from nuclear accidents and how liability and compensation regime have taken shape. The US Price-Anderson Act and the Paris Convention and its amendments, from the perspective of political economy, law and economics have a dominating presence in the writings. Academic work on the comparative legal jurisprudence and on the way the legal systems have evolved on the basis of the Paris and Vienna Conventions is very less.

On the requirement of a special liability regime for nuclear energy, Schwartz (2006, 2010) describes in clear terms that traditional tort law was seen as a major legal obstacle to the development of nuclear energy. According to her, "given the overwhelming complexities of task of identification of responsibility, tort law was not enough to facilitate victims for redress and compensation from a nuclear accident". Schwartz observes that the special regime that subsequently developed is the foundation for today's international legal regime on civil nuclear liability.

Though considerable progress has been made, many researchers have criticised the way in which international law relating to nuclear liability has developed, led by the Paris and Vienna Conventions. Van Dyke (2006) argues that there are "unresolved issues in the existing laws which demonstrate further work is needed to develop a comprehensive and authoritative regime to govern harm from nuclear activities". According to him, jurisdictional issues and neutrality of tribunal that are set differently by countries are major restrictions for victims of nuclear damage. Pelzer (2010a–c) has reviewed international nuclear liability laws from 1987 to 2004. Based on this study, although he agrees with Schwartz's view on successful development of the basic principles of liability laws as the landmarks, however, he argues that "only close to sixty states are either party to the liability conventions or have enacted national laws". He observes that there are sufficient nuclear liability regimes but still non-contracting States will revisit a 'Chernobyl situation'.

In the context of tort liability regime for transboundary accidents, Boyle (2005) clearly brings out the difficult nature of law making. He cites the International Law Commission's three decades of continuing effort in reaching a consensus on "Liability for Injurious Consequences of Acts Not Prohibited by International Law". Boyle in the context of sectoral agreements states that nuclear energy liability conventions represent an alternative approach to transboundary liability, which has had some success.

In the general context of international environmental agreements, Sachs (2008) is of the view that even after bold initiatives, "several civil liability treaties have been negotiated only a few have entered into force, and most remain", as she puts it, "unadopted orphans in international environmental law". She points out that other than oil spills and nuclear accidents conventions, no civil liability treaties are in force. Sachs agrees that adherence to nuclear liability conventions have been better than other conventions, however, Sachs does not dwell much on the complexities of nuclear liability regime. Many authors have termed the nuclear liability regime as a very complex labyrinth.

For a variety of reasons, this complexity has arisen largely on account of fear of participating in the regimes. For nuclear energy countries, the worry is on account of accountability and State liability, and for non-nuclear countries the fear is due to the perceived bias. Currie (2008) criticises international nuclear liability regime as extremely patchy and complicated, which features sparse participation. Though he welcomes the amendments to the Vienna and Paris Conventions, he opines that "they are heavily hedged with exceptions and the amended Protocols enjoy even more sparse participation than the original Conventions". And in terms of who should bear the liability, backed by IAEA explanatory notes, he observes, "channeling benefits the nuclear industry and its suppliers, as it focuses liability on one party who can then insure, but it prejudices the victim as it limits the parties against whom they may claim. Channeling of liability to one party needs rethinking, and must be borne by the parties involved, who should bear joint and several liability". The IAEA Explanatory Text (2007a, b) states:

> Like the principle of strict liability, the principle of exclusive liability of the operator facilitates the bringing of claims on the part of the victims of a nuclear incident, since it relieves them of the burden of proving the liability of parties other than the operator. But the principle also obviously favours the manufacturer, supplier or carrier of the material or equipment, since it obviates the necessity for them to take out insurance, as well as any other person who may have contributed to the nuclear incident.

The backgrounds of these articles and of several others are very compelling. Unlike other accidents, a nuclear incident covering wide geographical area with widespread impact has effects that are unquantifiable and long term. It is therefore imperative for the international community, to not just draft laws, but to bring into its fold as many countries as possible. Basically, foundational principles of civil nuclear liability must be incorporated in all the domestic nuclear laws; either through adherence, enacting national laws or in any other way. As Sachs (2008) observes, civil liability rules should not be like the "Yeti of international environmental law—pursued for years, sometimes spotted in rough outlines, remarkably elusive in practice".

The CSC negotiated under the auspices of IAEA (1997) is the latest convention to attempt further universalisation of the liability regime. McRae (2007) argues CSC can "become the catalyst for a Global Nuclear Liability Regime—having greater legal certainty as well as assuring meaningful compensation with a minimum of litigation". This argument may have some merits, but global adherence is still far away.

The fact that, even today, we are discussing issues relating to cross-border nuclear liability shows that the development of an acceptable regime is far from adequate. Several studies from Europe point out the continuing nature of the 'patchwork' liability law of the EU States. One must understand the importance of the EU as a region in this context. Europe is the region which has had the effect of the Chernobyl accident. EU, unlike other large nuclear power producing countries, is a constellation of States bordering each other and an accident in one State is bound to have major repercussions in many other countries. The European Commission (2007) has initiated further attempts to harmonise several EU country laws.

Although from the European Context, Pelzer's (2010a–c) argument on creation of nuclear-risk community seems to be the next logical step. He concludes that, "even if one accepts that a globally unified liability regime most probably is not achievable and that adherence to the international regime by a number of States is 'not even wanted' there is nevertheless the urgent need to strive for harmonisation in larger geographical regions which qualify for establishing risk community". Further, they see the recent developments in the European Commission (EC) preparing for the establishment of a unified nuclear liability regime within EU member States as most desirable. In fact, some of the recent works from Europe (Handrlica 2010) are advocating further EU involvement in the area of nuclear liability by means of the European Nuclear Liability Directive.

While Europe goes through a consolidation phase, US and Canada offer a different structure in respect of transboundary legal reach, harmonisation, compensation, etc. In respect of transboundary legal reach, the Canadian liability law states that an operator is not liable for any injury or damage occasioned outside Canada unless a country is declared as a reciprocating country. The Rules allow them to implement any arrangement between Canada and the reciprocating country, and injury or damage incurred in that country may be treated as similar to an injury or damage incurred in Canada. *The Canada-United States Nuclear Liability Rules (C.R.C., c. 1240)*[8] enacted under the provisions of Nuclear Liability Law recognises the US as a reciprocating country. The Price-Anderson Act defines 'nuclear incident', as "an incident occurring inside the US causing injury or damage inside or outside the US".

The Price-Anderson Act defines "nuclear incident" as an incident occurring inside the US causing injury or damage inside or outside the US. This could mean nuclear incidents within the United States causing damage within and outside the United States. By this the coverage provided by the act may apply to Canadians suing in a US court. According to Roman et al. (*n.d.*) this question was considered by the Nuclear Regulatory Commission (NRC) while interpreting the Atomic Energy Act of 1954. NRC concluded that there is nothing limiting compensation for damage incurred outside the US. The position as of today in both Canadian and US regimes is explicit in providing compensation as a reciprocal measure.

With respect to liability law developments in specific countries, Lee (2000) provides a critical account of the principles operating in civil nuclear industry in

[8] Nuclear Liability Act, "Declaring the United States to be a Reciprocating Country for Purposes of the Act (SI/78-179)".

the UK. She is of the view that international liability regimes may look good on the surface but are inherently faulty. While examining how the Paris Convention principles are applied in the UK, she argues that "there are real obligations to satisfy the interests of a range of environmental, property and personal interest which might be affected by nuclear operations". Her analysis is of great relevance to the wider debate on environmental liability regimes.

Many developments have happened with respect to liability law in the UK both in terms of domestic efforts and in the broader adherence to the Paris Convention and its protocols. The nuclear industry is an area where liability regimes are governed by statutory provisions and not by the strict liability rule expounded by *Ryland v. Fletcher*. The basic legislation on nuclear third party liability in the UK is contained in the *Nuclear Installations Act (NIA) 1965*, which implements the provisions of the 1960 Paris Convention and the 1963 Brussels Convention. NIA was amended in 1983 to give effect to two 1982 Protocols. Interestingly, substantial changes to NIA 1965 are now being considered. This follows from the release of the government's public consultation 'Implementation of changes to the Paris and Brussels Conventions on nuclear third party liability' in January 2011[9] (Department of Energy and Climate Change, United Kingdom 2012). One of the major changes is on transboundary jurisdiction. Presently, only claimants in a Paris Convention state can claim for damage arising from a nuclear incident. Under the new amendments operators will be liable to non-Paris Convention States under certain conditions.

Remarkably, within Europe, Austria has defied the conventional liability principles and enacted the *Third Party Nuclear Liability Act* in 1998. The law offers unlimited liability; no-legal channeling; and the power to Austrian courts to consider the extent of compensation and to determine whom to compensate, etc. The researcher views these as a clear indication that even OECD countries which had the benefit of a regional framework for long, are considering substantial changes to achieve a harmonised regime.

An important milestone from the perspective of developing countries is India's nuclear liability law. India, after nearly three decades of international nuclear isolation, became a part of the nuclear community subsequent to the *India–United*

[9] Public consultation aims to seek the view on the UK's proposed implementation of amendments to the Paris Convention on nuclear third party liability and Brussels Supplementary Convention. The main changes proposed are in three areas:

1. increase in the categories of damage for which operators are liable including damage related to the environment
2. widening the geographical scope of those that are eligible to claim compensation
3. a significant increase in the financial liability of the operator from currently £140 million to €1200 million

The Government response to the consultation has been published. As per the information, the Government of UK intends to implement its proposals as set out in the consultation paper, taking account of the responses submitted (All the documents including the responses from public and others are available in the website of Department of Energy and Climate Change, UK).

States Civil Nuclear Energy Cooperation Agreement in 2005[10] and later became eligible for international nuclear trade and commerce when Nuclear Suppliers Group[11] (NSG) granted wavier to India in 2008. Among several legal measures required to initiate the import of nuclear reactors from other countries, liability law is considered as one of the crucial legal elements. After much deliberation and debate, the Indian Parliament enacted *Civil Liability for Nuclear Damage Act 2010*. In October 2010, India signed the CSC but is yet to ratify. The Indian nuclear liability law, its content and the process of law making have been unique in many ways and as a test case for several of the international legal scenarios from an emerging country.

Before the legislative enactment of the nuclear civil liability law, India developed a strong tort liability system based on a large body of judicial pronouncements. While the House of Lords in the UK was reluctant to enlarge the concept of strict liability, the Indian courts, particularly the Supreme Court, expanded and innovated strict liability principles. Absolute liability, with no exceptions to liability, unlike the principles in strict liability became part of Indian law through the decision of the Supreme Court in *M C Mehta v. Union of India, AIR 1987 SC 1086* (Oleum Gas Leak Case).

Several authors have written on the political, legal and judicial experience of the Bhopal accident and the subsequent progress of liability law in India. It may be remembered that before the Supreme Court took a view on the Bhopal gas leak case in 1986 while deciding the *Oleum Gas Leak Case*, the Court had expanded the meaning of principle of liability and said:

> We are of the view that an enterprise which is engaged in a hazardous or inherently dangerous industry which poses a potential threat to the health and safety of the persons working in the factory and residing in the surrounding areas owes an absolute and non-delegable duty to the community to ensure that no harm results to anyone on account of hazardous or inherently dangerous nature of the activity which it has undertaken.

The Indian nuclear liability law, enacted nearly three decades after the Bhopal gas tragedy, is an attempt to address statutorily, the specific issue of nuclear liability and compensation.

With the legislative enactment of the civil nuclear liability law, India hopes to expand the nuclear power programme considerably by addressing the concerns of the public with sufficient mitigating measures, and that of the industry by limiting the liability both in terms of amount and time. Along with the legislative process, India became signatory to the CSC, thus becoming a part of an international liability regime. There have been numerous writings on the content of the law and

[10] Agreement signed by India and United States on July 18, 2005 towards U S help in removing three-decade U S moratorium on nuclear trade with India. The U S Congress on October 1, 2008, gave final approval to an agreement facilitating nuclear cooperation between the United States and India.

[11] The NSG is a group of 44 nuclear supplier countries, and works as a nuclear cartel. It seeks to contribute to non-proliferation of nuclear weapons through implementation of guidelines for nuclear exports and nuclear-related dual use items.

how it is different from other national laws, and whether the law is in variance with the international commitment. Deviations were particularly in three aspects, on the limitation in liability, legal channeling and on the right to recourse. Though there has been some discussion on India's divergent stance on CSC, strict provisions relating to liability claims and right of recourse were retained. It was also in parts argued that India's attempt should be considered as the next step in the evolution of nuclear liability law. While analysing the Indian law that was tabled in the Parliament for discussion, Hariharan (2011) observes that even though India faced a lot of criticisms in changing the rules of nuclear liability game, the Indian bill provides an opportunity to modernise the out-dated nuclear liability regime. She further observes that even with few interpretational issues, the Bill "is a significant step towards providing the public and the environment with greater protection by updating the nuclear liability regime to reflect the industry's current status as a mature industry no longer in need of such strong liability protections". The final liability law went much ahead on the questions of liability amount and claims and also on the operator's right of recourse against suppliers. Although much has been talked on the substantive content of the law, there has not been any discussion either in public or in Parliament on the harmonisation issues in the liability framework that India and the rest of region may have to face.

Internationally, however, the Fukushima nuclear accident in Japan on 11 March 2011 substantially altered the discourse on the nuclear liability regime. IAEA was again faced with the question of how to universalise and establish a truly global nuclear liability regime. The IAEA Ministerial Conference on Nuclear Safety (2011a–c) in the aftermath of the Fukushima accident concluded with the adoption of a declaration by the IAEA member States. On nuclear liability, the declaration states:

> Recognise the need for a global nuclear liability regime that addresses the concerns of all States that might be affected by a nuclear accident with a view to providing appropriate compensation for nuclear damage.

The Fukushima accident being recent, Pelzer (2011) reminds us that "States used the momentum of the 1986 Chernobyl accident in a most effective way and question whether the world community would seek and push for a comparable momentum consequential to the Fukushima nuclear incident". Considering the worldwide reaction in the aftermath of Fukushima, Reyners (2013), observes that the accident in Fukushima has generated an emotion not experienced since Chernobyl 25 years ago. He questions "whether and how this emotion could transform itself in a real progress for the global regime of nuclear safety?" This is a question that needs further deliberation. A scenario does exist very similar to the events after the Chernobyl accident and the IAEA (2011a–c) re-emphasising the desire to forge consensus, points towards a risk precaution in larger geographical areas. In summary, the existing literature in this field mostly points out to the fact that work on a universal law is far from over. Experts doubt the possibility of such a law that has universal acceptance. Further, at present there is no discussion on the transboundary risk or legal implications arising from the nuclear energy programmes in South Asia.

Many authors argued the existence of an international patchwork regime; however, not many reasons were brought out on the continuing reasons for the same. Though there is a movement towards a regional agreements led by the EU and, reciprocity between Canada and US, this mechanism as a way forward to other regions have not been given adequate attention.

1.6 Objectives and Scope

The objective of the study is to:

1. Critically study the laws relating to transboundary nuclear liability at the regional and international level.
2. Test the compatibility of Indian nuclear liability law *vis-a-vis* Convention on Supplementary Compensation in respect of transboundary application of liability.
3. Analyse nuclear energy programmes, international obligations and SAARC structure in South Asia to find out how the region is placed in responding to a transboundary nuclear impact.
4. Determine through technical study whether South Asia can be considered as a nuclear-risk region, and examine a possible institutional arrangement based on the views of experts.

The study is purely in the context of transboundary risk and its legal consequences emanating from a civilian nuclear power plant. The work examined the historical advancement of nuclear technology from military to civilian applications and developments leading to the internationalisation of nuclear liability laws. The study examined the Paris Convention, the Vienna Convention, CSC, and also the domestic laws of select jurisdictions such as the US, UK, India and Bangladesh. Selections of these countries are based on several factors. The UK is a common law country with a domestic liability law and is also a party to the Paris Convention. The US has a unique domestic legislation.

The Fukushima nuclear accident occurred during the course of this study. The Chernobyl and Fukushima accidents are utilised to identify the geographical scale and impact of a nuclear disaster. The purpose is to extrapolate this information into the South Asian context and to find out the risk probability in the region.

With respect to the risk assessment study, the whole of South Asia (eight countries in SAARC) is taken into consideration, both for technical risk-zone assessment and for expert interviews. The rationale behind the technical study is to assess the probable risks that exist with the development of nuclear energy. With the technical results in hand, interviews with experts from SAARC countries were undertaken to get first hand information on the level of perceived risk and the possible ways of addressing the risk within South Asia.

Since the research is from India, a critical review of the Indian nuclear liability law is undertaken to study the extent of convergence and divergence with the Convention on Supplementary Compensation in respect of transboundary liability application.

The study does not include accidents which are domestic, i.e. without any transboundary consequences. For instance, in the Three-Mile accident, the release of radiation was confined to the US. The study is strictly confined to the realm of the civil nuclear liability regime and did not undertake criminal, economic and suppliers' liability. The scope of the study is limited to the transboundary nature of the nuclear disaster from a civilian nuclear power project. Military facilities are outside the purview of the study. In respect of interview, interview with Pakistani officials could not be undertaken.

1.7 Methodology

A comprehensive literature survey was done with respect to all relevant international treaties and conventions and the laws of select countries. The primary and secondary literature on adoption and explanatory text of conventions, domestic legislations and critical writings formed the basis of the study. IAEA responses to the Chernobyl and Fukushima were considered as the basis of reform of liability regime. With respect to South Asia, a detailed study and analysis were undertaken on the nuclear energy programmes and plans; international nuclear legal obligations; domestic nuclear energy laws and policies, and disaster management initiatives. Documents on SAARC were collected from SAARC Documentation Centre, SAARC Disaster Management Centre and SAARC Law Centre.

In order to determine and plot the nuclear-risk zones, Geographical Information System (GIS) software (ESRI's ArcGIS Version 10) was used. Risk zone data from IAEA Safety Series, US NRC, Atomic Energy Regulatory Board (AERB) and EC study on nuclear emergency zoning, and also of the UN and other non-governmental studies on Chernobyl and Fukushima impact were relied upon. The zones that were taken into considerations are 30, 50, 100, 300, 500 and 1000 km. Exact locations of the all (current, under construction and planned) NPPs were taken through measurement of latitude and longitude using Google maps.

NPP locations were sourced from government data. Wherever there is no official communication, especially in respect of new proposals, reported news was relied upon. In the case of Sri Lanka, since there has not been any official announcement of nuclear plant sites—two hypothetical locations were selected.

The technical results were followed up through interviews to understand how risk is perceived by policy makers and the nuclear community. For interview, the qualitative method was undertaken. The interviewees were from the nuclear scientific community; political counselors from SAARC embassies; SAARC institutions; academicians; lawyers; diplomats; science reporters; disaster management experts and also international nuclear lawyers.

References

Boyle, A. E. (2005). Globalising environmental liability: The interplay of national and international law. *Journal of Environmental Law, 17*(1), 3–26.

Commission of the European Communities. (2007). *Nuclear illustrative programme: Presented under Article 40 of the Euratom treaty for the opinion of the European economic and social committee.* Retrieved March 14, 2011, from http://eurlex.europa.eu/LexUriServ/LexUriServ.do?uri=COM:2006:0844:FIN:EN:PDF.

Currie, D. E. J. (2008). The problems and gaps in the nuclear liability conventions and an analysis of how an actual claim would be brought under the current existing treaty regime in the event of a nuclear accident. *Denver Journal of International Law and Policy, 35*(1), 85–127.

Department of Energy & Climate Change, UK. (2012). *Implementation of changes to the Paris and Brussels conventions on nuclear third party liability a public consultation.* Department of Energy & Climate Change, UK. Retrieved September 5, 2012, from http://www.decc.gov.uk/en/content/cms/consultations/paris_brussels/paris_brussels.aspx.

Handrlica, J. (2010). Euratom powers in the field of nuclear liability revisited. *International Journal of Nuclear Law, 3*(1), 1–18.

Hariharan, A. (2011). India's nuclear civil liability bill and supplier's liability: One step towards modernizing the outdated international nuclear liability regime. *William & Mary Environmental Law and Policy Review, 36*(1), 223–255.

IAEA. (1997). *The 1997 Vienna convention on civil nuclear liability for nuclear damage and 1997 convention on supplementary convention for nuclear damage-explanatory texts.* Retrieved March 20, 2010, from http://wwwpub.iaea.org/MTCD/publications/PDF/Pub1279_web.pdf.

IAEA. (2007a). *Energy, electricity and nuclear power estimates for the period up to 2030*, Vienna, IAEA-RDS-1/27.

IAEA. (2007b). *Energy, electricity and nuclear power: Developments over 25 years and future projections*, Vienna, STI/PUB/1304.

IAEA. (2011a). *Draft IAEA action plan on nuclear safety.* IAEA GOV/2011/59-GC(55)/14, 5 September. Retrieved March 12, 2012, from http://www.iaea.org/About/Policy/GC/GC55/Documents/gc55-14.pdf.

IAEA. (2011b). IAEA Annual Report.

IAEA. (2011c). Declaration by the IAEA ministerial conference on nuclear safety in Vienna on 20 June 2011, INFCIRC/821.

IAEA. (2012). Progress in the implementation of the IAEA action plan on nuclear safety, supplementary information. General Distribution, (ref: GOV/INF/2012/11-GC(56)/INF/5), 15 August. Retrieved August 28, 2012, from http://www.iaea.org/About/Policy/GC/GC56/GC56 InfDocuments/English/gc56inf-5-att1_en.pdf.

Lee, M. (2000). Civil liability of the nuclear industry. *Journal of Environmental Law, 12*(3), 317–332.

McRae, B. (2007). The convention on supplementary compensation for nuclear damage: Catalyst for a global nuclear liability regime. *Nuclear Law Bulletin, 79*(1), 17–36.

Pelzer, N. (2010a). Main features of the revised international regime governing liability-progress and standstill. In *International nuclear law: History, evolution and outlook.* Paris: OECD.

Pelzer, N. (2010b). Compensation for large-scale and catastrophic nuclear damage. In Nótári, T. & Török, G. (Eds.), *Unnepitanulmanyok: Lamm Vanda tiszteletere* (p. 344). Budapest: JogtudomanyiIntezete.

Pelzer, N. (2010c). Main features of the revised international regime governing liability-progress and standstill. In *International nuclear law: History, evolution and outlook* (pp. 355–386). Paris: OECD/NEA.

Pelzer, N. (2011). Does the Fukushima nuclear incident require a revision of the international legal regime on nuclear safety? International Atomic Energy Agency. Retrieved October 5, 2012, from http://www-pub.iaea.org/mtcd/meetings/pdfplus/2011/cn200/working-sessions/w_d4_n.pelzer.pdf.

Reyners, P. (2013). A new world governance for nuclear safety after Fukushima? *International Journal of Nuclear Law, 4*(1), 63–77.

Sachs, N. (2008). Beyond the liability wall: Strengthening tort remedies in international environmental law. *UCLA Law Review*, *55*, 837, 838–861.

Schwartz, A. J. (2006). International nuclear third party liability law: The response to chernobyl. In *The international nuclear law in the post-chernobyl period*. Paris: OECD.

Schwartz, A. J. (2010). Liability and compensation for third party damage resulting from a nuclear accident. In *International nuclear law: History, evolution and outlook* (pp. 307–354). OECD/NEA.

The Price-Anderson Nuclear Industries Indemnity Act. (1957). United States Code Annotated § 2210.

Van Dyke, J. M. (2006). Liability and compensation for harm caused by nuclear activities. *Denver Journal of International Law and Policy*, *35*, 13–46.

Chapter 2
The Development of Institutions and Liability Laws Relating to Nuclear Energy

Abstract Nuclear energy development and the laws relating to liability and compensation have had an eventful history, which continues even today. Historically, nuclear energy development was initiated as a military programme, and once the war ended, efforts were made to use this energy source for civilian applications. Today, world over, institutions have been established to control and propagate the use of civilian nuclear energy in line with internationally accepted rules. Over the years, however, the memory of atomic bombings in Japan, accident at Three Mile Island, Chernobyl and recently as Fukushima has left the public with a deep suspicion of the civilian use of nuclear energy. Laws relating to liability and compensation in case of accident became the foundation to take the program forward. In order to create a universal liability regime, regional Paris Convention and international Vienna Convention formed the template. Many countries followed these conventions based on geographies with varying degrees of adherence, while several countries remained outside the regime. It appears that even after many decades of the Chernobyl transboundary nuclear accident and the recent Fukushima accident, a universal regime is far from being realised.

Keywords International atomic energy agency · Nuclear energy and technology transfer · Bilateral agreement on nuclear energy · Nuclear risk studies · Military to civilian nuclear program · Development of nuclear liability regime · OECD and United States nuclear liability law

2.1 Introduction

The development of nuclear energy for peaceful purposes has its history in military applications. In the aftermath of the horrific bombings of cities in Japan in 1945, the World community, primarily led by the United States (US) and the

M.P. Ram Mohan, *Nuclear Energy and Liability in South Asia*,
DOI 10.1007/978-81-322-2343-6_2

Soviet Union (USSR), insisted on regulating nuclear technology, convinced that such regulations should not hamper the progress in peaceful applications. Countries saw the enormous potential of nuclear technology in its civilian use, particularly in nuclear power programs.

With the end of the war, efforts were redirected to produce civilian applications for nuclear technology (Green 1973). Governments actively pursued domestic nuclear energy programs and also entered into bilateral agreements for transfer of technology. Initially however, such transfer of technology was limited to war-time allies. Many countries outside the military alliance sought technology for their power programs in the hope that this would give them unlimited source of power and would enable them to economically progress faster (Fischer 1997). This global interest called for the creation of an international institution at the United Nations (UN), tasked with regulating the security and safety of nuclear technology. In subsequent years, international community established the UN Atomic Energy Commission (UNAEC) in 1946 and later in 1957 International Atomic Energy Agency (IAEA) came into existence.

Pursuing civilian programs also meant setting up of adequate mechanisms for tackling the consequences of an accident. The impact could lie within the country or be transboundary in nature. The problems of civil liability for nuclear damage became one of the most contested issues. Knowing well that nuclear activities are more hazardous than conventional industrial activities, the need for a specific legal regime for nuclear damage was felt (Schwartz 2006). It was argued that this would ensure that countries could provide prompt and adequate compensation for nuclear damage without exposing the infant nuclear industry to excessive burdens (IAEA 1997).

In 1986, the Chernobyl accident in former USSR revealed the extent of damage that can be caused by a nuclear accident. Radiation transgressed national boundaries and caused harm on an international scale (IAEA 2006). Much before Chernobyl, other nuclear accidents, 1957 Windscale in the United Kingdom (UK) and 1974 Three Mile Island in the United States had also occurred. But the damage from both Windscale and Three Mile Island was limited to reactor surroundings and there were no transnational implications. The Chernobyl accident shook the international community; subsequently major reforms were undertaken to strengthen the international nuclear liability regime. Unfortunately, it took another accident; the 2011 Fukushima nuclear accident in Japan, to reinforce the importance of a universal regime which is in harmony within the wider regions of the world.

This chapter briefly describes the historical facts relating to the development of nuclear technology. This is followed by a discussion on the civilian applications and consequent development of liability regimes. The special legal regime sought and enacted; and post-Chernobyl modernization attempts are dealt in detail. The chapter concludes with an analysis of these conventions in order to grasp the difficultly in achieving a global nuclear liability law.

2.2 Institutional Control of Nuclear Energy: Creation of the UN Atomic Energy Commission

In July 1945, the US tested its first atomic bomb in Los Alamos, New Mexico. The test came nearly 3 years after Enrico Fermi's effort of achieving controlled nuclear chain reaction (US DOE 2006). Soon after, in August, the US atomic bombing of Hiroshima (6 August 1945) and Nagasaki (9 August 1945) in Japan showed the destructive power of nuclear energy, and the consequences of human exposure to nuclear radiation. It may be remembered that the action of bombing Japan took place within weeks of signing the Charter of the UN on 26 June 1945 in San Francisco (UN Charter 1945). With the war ending, and knowing well that the US may no more hold monopoly in nuclear science, efforts were directed towards controlling the spread of nuclear weapons (Green 1973). At the same time, attention was also focused on the peaceful application of nuclear energy. The newly formed United Nations was drawn into this effort.

UN became the focal point for nuclear issues and concerns. It was believed, as the subject is contentious, that the involvement of all the five permanent members of the Security Council and others at a multi-lateral institutional level was necessary (Fischer 1997). In 1945, there had been extensive consultations, in particular, between the US, the UK, Canada and later with the USSR on "the need in an effort to reach agreement on the conditions under which international co-operation might replace rivalry in the field of atomic power" (Bathurst 1947). In November 1945 at a multilateral summit in Washington; US, UK and Canada issued a declaration that, "in order to attain the most effective means of eliminating entirely the use of atomic energy for destructive purposes and promoting its widest use for industrial-and humanitarian purposes, a commission should be set up at the earliest practicable date, under the UN, to prepare recommendations for submission to that Organization" (Bathurst 1947).

On 27 December 1945, at a meeting of Council of Foreign Ministers held in Moscow, the creation of the United Nations Atomic Energy Commission (UNAEC) was proposed by the US and UK, and agreed to by USSR with veto conditions. On 24 January 1946, in its first Session, the UN General Assembly created the UNAEC (UN Resolution 1946). The Commission was entrusted to proceed "with the utmost despatch" and towards an effort "to consider problems arising from the discovery of atomic energy and related matters" (Fischer 1997). The terms of reference were:

> The exchange of basic scientific information for peaceful use between all nations;
> The control of atomic energy to the extent necessary to ensure its use only for peaceful purposes;
> The elimination of national armaments adaptable for mass destruction, atomic or other;
> Effective safeguards by inspection or other means to protect against the hazards of violation or evasion.

However, though well intended in its efforts in establishing an international oversight on use of atomic energy, the working of the UNAEC had been contentious. The US presented the Baruch Plan in March 1946 and the USSR presented the Gromyko Plan in June 1946; both the proposals had different visions on the structure and mission of the UNAEC. It is reported that not a single unanimous decision by the Veto powers, particularly the US and the USSR could be achieved (Fischer 1997). While reviewing the ongoing work at the UNAEC, Benoit-Smullyan (1946) summarised that the heart of failure of negotiations on atomic bomb was the 'distrust' between the US and the USSR. The work of UNAEC suffered greatly; the Commission discontinued its working even without being discharged by the Security Council (Shils 1948). In September 1949, the USSR conducted its first nuclear weapons test followed by the UK in 1952 (Fischer 1997). The Soviet nuclear test effectively ended the UNAEC's role, and in 1952 the UN General Assembly formally dissolved the UNAEC. A view may be taken that the entire negotiating exercise was undertaken superficially with no intent to broke a conclusive deal, but only to show deliberation as a cover for building their nuclear capabilities. The lack of any international institutional space for deliberation after the dissolution of the UNAEC led each country to pursue its own strategy on nuclear energy.

In the meanwhile, an important step towards peaceful uses of nuclear energy—the production of electricity from atomic energy became successful. In 1951, electricity generation was successful (100 kilowatts (KW) of electricity) from an Experimental Breeder Reactor (EBR-I) at the National Reactor Testing Station in Idaho, the US (Michal 2001). Even before this success, after the Second World War with the intent to institutionalise the nuclear energy program, the US government enacted the Atomic Energy Act on 1 August, 1946 (US AE Act 1946).

One of the major aims of the law was to harness the use of atomic energy for peaceful purposes. The US AE Act created the United States Atomic Energy Commission (USAEC) as a civilian controlled authority. Consequent to the establishment of the USAEC, the complete Manhattan Project office and its related facilities and files were transferred to the USAEC (Buck 1983). Acknowledging the importance of civilian applications, the US Congress while enacting the US AE Act stated that "atomic energy should be employed not only in the Nation's defense, but also to promote world peace, improve public welfare, and strengthen free competition in private enterprise" (Buck 1983). However, the US AE Act retained complete government control, albeit with exceptions, of the nuclear materials—its production, research and use.

Meanwhile in USSR, the 1940s and 1950s witnessed considerable effort in developing nuclear technology (both for military and civilian purposes) with the aim to reach the same capability as the US (Nordykea 1998). The Institute of Physics and Power Engineering was set up in May 1946 at the then-closed city of Obninsk, 100 km southwest of Moscow, to develop nuclear power technology. Obninsk became one of the nerve centres of the Soviet nuclear energy program and became famous on 27 June 1954 with the world's first nuclear power station at Obninsk with a capacity of 5 MW, being connected to the Moscow grid

The US by walking the talk at the UN, in 1954, provided a legal basis for 'Atoms for Peace' by amending its Atomic Energy Act, 1946. This led to a series of international nuclear cooperation, concluding as bilateral agreements with a number of States (Fischer 1997). The amended Atomic Energy Act, 1954 (US AE Act 1954) drastically altered the policy, thereby facilitating participation of private enterprises in the development of atomic energy (Sect. 3, Atomic Energy Act 1954).

2.2.1 Military to Civilian Programs: Creation of International Atomic Energy Agency

Atoms for Peace program gave a foundation to collaborate, and thereby transfer the technology for peaceful applications. In parallel, efforts to establish IAEA envisaged in the 'Atoms for Peace' program also gathered pace. The UN General Assembly in its plenary session, in December 1954, unanimously adopted a resolution for the establishment of the IAEA, and for the convening of an international technical conference of governments under the auspices of the UN (IAEA 1964).

The *International Conference on the Peaceful Uses of Atomic Energy* was convened under the auspices of the UN in Geneva, in August 1955. The event was a landmark inter-governmental conference, held to illuminate to the world the progress in nuclear science and technology, and it also demonstrated that the generation of electricity was possible (Krige 2006). The President of the Conference, Homi Bhabha, in the context of future scientific and technological progress envisioned that, "during the next two decades scientists would have found a way of liberating [thermonuclear] fusion energy in a controlled manner... When that happens the energy problems of the world will truly have been solved forever..." (Fischer 1997).

For the international institutional initiative, the success of the international conference took the negotiations in establishing IAEA to a level of urgency. From 1955 to 1957, after a series of negotiations, 12 governmental representatives from Australia, Belgium, Brazil, Canada, Czechoslovakia France, India, Portugal, South Africa, the UK, the US and the USSR concluded the drafting of the IAEA Statute (Fischer 1997). India has been a member of the Board of Governors of the IAEA since its inception. Chidambaram (2006) says that the Indian contribution during the negotiations was in the structuring of the IAEA Governing Board. He explains "the Indian delegation came up with a complex but ingenious formula for determining the composition of the IAEA Board of Governors". The Statute of the IAEA was approved on 23 October 1956 by the Conference on the Statute of the IAEA, which was held at the Headquarters of the UN. It came into force on 29 July 1957. The establishment of IAEA was with a twofold purpose: to promote the peaceful uses of atomic energy, and to ensure, so far as it is able, that such uses do not further any military purpose (Willrich 1965). The main functions that the IAEA was to perform are (Article III, IAEA Statute):

1. Take any action needed to promote research on, development of, and practical applications of nuclear energy for peaceful purposes (Article III.A.1);
2. Provide materials, services, equipment and facilities for such research and development, and for practical applications of atomic energy "with due consideration for the needs of the under-developed areas of the world" (Article III.A.2);
3. Foster the exchange of scientific and technical information (Article III.A.3);
4. Encourage the exchange of training of scientists and experts in the field of peaceful uses of atomic energy (Article III.A.4);
5. Establish and apply safeguards to ensure that any nuclear assistance or supplies with which the IAEA was associated should not be used to further any military purposes—and apply such safeguards, if so requested, to any bilateral or multilateral arrangement (Article III.A.5);
6. Establish or adopt nuclear safety standards (Article III.A.6);
7. Acquire or establish any facilities, plant, and equipment useful in carrying out its authorised functions (Article III.A.7).

Bechhoefer and Stein (1957) summarises the core functions of the IAEA into two: "positive function and negative function". With the positive function being to "seeks to accelerate and enlarge the contribution of atomic energy to peace, health and prosperity throughout the world" and the negative function "insuring, so far as it is able, that assistance provided by it or at its request or under its supervision or control, is not used in such a way as to further any military purpose." In order to carry forward the vision, i.e. expanding peaceful applications of nuclear programs around the world and securing its safety, the immediate tasks of IAEA were to frame rules of safeguards.

Importantly for the purpose of this study, countries, having known the destructive power of nuclear energy from the nuclear bombings in Japan, were becoming concerned about the liability and the accepted definitions of liability arising from an accident in a civilian nuclear reactor. They argued that without a clear understanding of the State responsibility on nuclear liability, growth of the atomic energy industry remains in doubt.

2.3 Nuclear Energy Cooperation: Questions of Liability

In order to achieve the vision envisaged in the 'Atoms for Peace' program, i.e. advancing international cooperation; bilateral agreements were negotiated by technology supplier countries with friendly countries (Mallison 1957). There were numerous complex problems that arose while negotiating these agreements. One such complex issue is the State responsibility and its liability.

Any nuclear operation involves use of radioactive materials. Radioactive materials are hazardous in nature and require careful management. The most significant radioactive hazards come from a catastrophic reactor accident. Even though Edwards (1957) considers such a possibility extremely rare, but "given the worst possible combinations of circumstances that might conceivably occur—combinations of mechanical and human failure and error and weather conditions—there exists the possibility of a catastrophic incident causing personal injury and property damage of unprecedented magnitude". The nuclear bombings in Japan set the context of what constitutes risk in the event of a nuclear disaster.

2.3.1 Technical Studies on Hypothetical Nuclear Accidents

In order to develop nuclear energy on a commercial scale, by 1950s it became necessary to establish the scale of a probable accident, and to determine the amount of liability insurance that would be required to cover the cost of an accident. In March 1957, as a first attempt to estimate the possible consequences of an accident—the US Atomic Energy Commission (USAEC) published its risk assessment study titled *Theoretical Possibilities and Consequences of Major Accidents in Large Nuclear Power Plants: A Study of Possible Consequences of Certain Assumed Accidents, Theoretically Possible but Highly Improbable, were to Occur in Large Nuclear Power Plants*—commonly called Brookhaven Report or WASH-740 (USAEC 1957).

The Brookhaven study hypothetically analysed the extent of damages from a 500 MW reactor located about 50 km away from a city of 1,000,000 population. The results showed that the accident could cause deaths and injuries upto 3400 and 43,000 people, and also $7 billion as property damages. The predicted results varied widely, depending upon which set of assumptions was used (Otway 1974). At the same time, the report made it clear, that such a probability of an accident is extremely rare (USAEC WASH-740 1957). Almost in parallel to WASH 740, another study *Report on the possible effects on the surrounding population of an assumed release of fission products into the atmosphere from a 300 MW Enrico Fermi plant located at Lagoona beach, Michigan* were undertaken (USAEC APDA-120 1957). According to this report "in case of Enrico Fermi plant experiences a core meltdown and releases 50 % of its fission products, with weather conditions carrying radioactivity to the most heavily populated areas of Detroit, there will be an estimated 133,000 deaths, 181,000 immediate injuries and 245,000 long-term injuries" (USAEC APDA-120 1957 in Fuller 1975).

In 1965, an updated Brookhaven report that took into account the increased size of later-model reactors estimated dramatically higher numbers of 45,000 deaths and 100,000 injuries, and $17–$280 billion in property damage and long-term contamination of an area to the size of Pennsylvania (Anderson 1978). Anderson (1978) states that the updates were kept confidential and it was only after a threat through a Freedom of Information suit that the working documents of the 1965 the updates of WASH-740 were released. This report triggered considerable public concern about the safety of nuclear energy. It was said, *though the report was intended to calm fears by showing the rarity of such accidents, in reality, it created more alarm and distrust toward nuclear energy* (Nadar and Abbots 1979).

In 1974, the US Nuclear Regulatory Commission (USNRC) replaced the 1957 Brookhaven Report with a more rigorous study called "Reactor Safety Study" (Known as the Rasmussen Report or WASH 1400) (USNRC 1975). However, the Rasmussen Report has been controversial due to its reliability (US Congress 1984). In 1990, the USNRC published, "Severe Accident Risks: An Assessment for Five US Nuclear Power Plants"—NUREG-1150 (USNRC 1990). NUREG 1150 replaced all the earlier WASH 1957 and WASH 1975 reports (USNRC 1990).

These reports trigged a panic reaction from both the public as well the private industries. Knowing well the probable consequences would be far reaching on human health, environment and economic as witnessed in the Atomic Bombings in Japan, nuclear technology countries ring-fenced themselves while entering into a range of agreements for technology transfer. The concerns arose from both the liability arising from the technology supplied and also from transboundary consequences in the aftermath of a nuclear accident. Importantly, countries began to worry about State liability and its extent in the event of a possible atomic reactor accident abroad.

2.3.2 Continuing International Cooperation

However, internationally, the worry of State liability *per se* did not deter Canada, the US, the UK and the USSR from entering into bilateral agreements with a host of countries that sought nuclear energy for peaceful applications, particularly for electricity. Willrich (1965) states that in the initial years of the IAEA, since the US was unsuccessful in negotiating an international approach to technology transfer, it decided to proceed on a bilateral basis. The USSR also similarly entered into agreements with many countries. As of today, the US has 26 agreements in force for peaceful nuclear cooperation (EURATOM is being considered as a single agreement, consisting of 27 European countries) (US Government Accountability Office 2010). The USSR also had bilateral agreements with countries including Bangladesh, China, Egypt, Poland, Romania, India, Germany, Czech Republic, South Africa and others (Ginsburgs 1960). Canada, the UK and others also followed similar approach. It was under these bilateral agreements, especially in the initial years, that many nuclear 'have-nots' including countries in South Asia initiated their nuclear research program.

In the case of South Asian countries, India, Pakistan and Bangladesh benefited from these bilateral collaborations. Under the aegis of the 'Colombo Plan'; Canada provided research reactor technology and the US became the heavy water supplier for India (Canada–India CIRUS agreement 1956). The Colombo Plan was conceived as a technical assistance program for South Asian countries, with Canada supplying technical and scientific assistance and training the Indian staff

(Petros'yants 1984). Morokhov concluds that invaluable lessons had been drawn from the Obninsk reactor (IAEA 1968). According to him, the learnings were, "(a) Transformation of nuclear energy to electricity was proved to be practical, (b) Atomic power was sufficiently reliable and flexible in operation and fulfilled the requirements for utilization in an electrical network and (c) It was completely safe both for personnel in the plant as well as the surrounding population" (IAEA 1968).

Remarkably, in the Asian context, India was leading the efforts on nuclear energy. The history of India's nuclear quest is fascinating. As early as 1944, Homi Bhabha the founder of the Indian nuclear program proposed creation of Tata Institute of Fundamental Research (TIFR) (Chidambaram 2006). With the establishment of TIFR in the year 1945, India launched into research in nuclear technology and science. On 15 April 1948, the government enacted the Atomic Energy Act and established the Atomic Energy Commission of India. Chengappa (2000) says that Prime Minister Jawaharlal Nehru, stating his vision on the nuclear energy program, observed, "We must develop this atomic energy quite apart from war—indeed I think we must develop it for the purpose of using it for peaceful purposes…of course, if we are compelled as a nation to use it for other purposes, possibly no pious sentiments of any of us will stop the nation from using it that way". India immediately launched its nuclear energy program on a full scale.

In 1955, India began construction of a 1 MW research reactor with British assistance. In 1957, the first nuclear reactor in Asia—Apsara was dedicated to the nation (Chidambaram 2006). Sethna (1979) states that British assistance was limited to supply of fuel elements. The reactor and all the equipments were designed and built by Indians. Within a year, i.e. 1955 India was negotiating with Canada for 40 MW reactors. Under the "Atoms for Peace" program, the US supplied heavy water and the reactor was named the 'Canada-India Reactor, the US' otherwise called as CIRUS (Hunt 1977). CIRUS became operational in 1960. Pakistan, the only other country in South Asia also reached a bilateral agreement with Canada, which is discussed below.

As several countries were acquiring nuclear capability, Epstein (1977) observes "the incentives and disincentives for countries in going nuclear vary between a combination of military, political, and economic concerns and motivations at particular period in time". The World community was concerned about the consequences of the proliferation of the technology. The technology has duel applications—military and peaceful. On 8 December 1953, "Atoms for Peace" speech by the US President Eisenhower before the UN General Assembly called for the governments, principally involved in making joint contributions, to set up an IAEA under the UN (IAEA Transcript 1953). Eisenhower spoke on the peaceful use of the proposed atomic energy agency:

> The more important responsibility of this Atomic Energy Agency would be to devise methods whereby this fissionable material would be allocated to serve the peaceful pursuits of mankind. Experts would be mobilized to apply atomic energy to the needs of agriculture, medicine and other peaceful activities. A special purpose would be to provide abundant electrical energy in the power-starved areas of the world.

(Hunt 1977). Atomic Energy of Canada Ltd. (AECL), according to Hunt described the project as 'perhaps the largest single contribution any country has yet made' in the international atomic energy field. In 1959, Canada also entered into agreement with Pakistan for peaceful application of nuclear energy. In the following years, both India and Pakistan took lead, and entered into several bilateral agreements with the US, the UK, France and the USSR, and also enhanced their relationship with Canada. This international cooperation came to an end, for both India and Pakistan, after India tested its nuclear device in 1974.

Recently, Bangladesh and Sri Lanka have been engaged with IAEA and technology providers to initiate nuclear energy programs through bilateral agreements (See Chap. 4).

The development of the nuclear industry and its consolidation started during the pre-world war phase and was primarily intended for military purposes. The technology was such, that many allied countries collaborated with each other on science, technology, fuel, etc., for making the atomic bomb—*the Manhattan Project.* Similarly the USSR, through a network of communist states developed its competence. These relationships reflected the military alliances from the Second World War. The peaceful application, the energy programme was conceived from the military projects. As the civilian projects involved technology, equipments, fuels, expertise from various countries; industry structures became complex. This complexity became a factor in determining the consequences, to whom and to what extent one country is responsible on account of a nuclear accident in another country. Since a bilateral agreement is between two States; the nuclear technology supplier State was reluctant to singularly take on all the burden of the 'State responsibility and liability'.

2.3.3 Bilateral Liability Provisions and Transboundary Liability Concerns

It is clear that the importance of liability concerns was mainly dictated by one of the beneficiaries of the US bilateral agreements—the private industry. The US, through these bilateral agreements, enabled the private industry in the US to undertake an extensive commercial program abroad. Much before the establishment of the IAEA there were concerns on the operational realities of a nuclear power plant. Green (1973) observes that in the US, "the initial rush of enterprises (private) to get into nuclear energy where thwarted by hard economic and operational realities, i.e. profitability and liability". With the amendment of the US AE Act 1954, the industry was thrown open for private sector. The infant private sector was apprehensive of the insurability of nuclear business from unknown consequences, which may give rise to large liability claims (Meehan 2012). An interesting study by the Harvard Law School and Atomic Industrial Forum—*A Forum Report: International Problems of Financial Protection Against Nuclear Risks* in 1959 brought out the possible liability on American nuclear exporters with other contracting States (Cavers 1964).

In order to avoid being subjected to liability claims, Hardy (1961) in his comments on the US treaty with Germany and Euratom, states that "a number of the treaties that provide enriched fuels and power reactors also contain 'hold harmless' clauses for exonerating the US Government from possible liability". This was a conscious decision taken by the nuclear technology providers, for exempting themselves from possible host-country domestic law or litigation, in respect of liability and compensation. The UK, Canada and others also entered into bilateral agreements, but were fewer compared to the US and the USSR. It is unclear from the literature review whether such exclusionary clauses were also utilised by the USSR in bilateral agreements. A detailed study of the USSR bilateral agreements by Ginsburgs (1960) does not bring out any of the US-type hold harm-less clauses, but suggests that USSR was more liberal in terms of providing technology.

The issue of transboundary liability and state responsibility became important once many countries in Europe initiated their own nuclear power programs. Europe, unlike the US or Canada, is a closely knit constellation of States bordering one another. An accident in one country is bound to have its consequential effect in other countries. Realising the problem of transboundary nuclear damage, Organization for European Economic Cooperation (OEEC)—an organisation established in 1948 for post-war reconstruction efforts; as early as in 1957, established a Working Group on Harmonization of Legislation to examine third-party liability for damage caused by the peaceful use of nuclear energy (Marcus 2008). International cooperation and standard settings became imperative due to the nature of the nuclear energy industry. Freeman (1960) rightly puts it "The concern is universal not merely because atomic energy will permeate-if it has not already done so-every facet of human endeavor, but also because the nuclear genie presents problems and risks which cannot and will not be solved satisfactorily on a purely national or local level. Both because of these risks and the resources required, the need for co-operative action, whether on a regional or a worldwide basis, is probably more insistent here than in other sectors of commerce and industry". Transboundary liability is one of the most contested issues and also one where there has been major international cooperation.

2.4 Specific Legislative Response on Nuclear Liability

From the early days of development of nuclear energy, it was recognised that sufficient protection must be accorded to the public against the probable risks arising from nuclear energy production. Equally significant is the risk borne by the nuclear industry—unresolved insurance and possible massive litigation for claims on account of an accident.

In many jurisdictions, tort rules govern liabilities arising from hazardous activity. These tort rules require proof of fault and causation. Additionally, any entity or person and any number of them can be held liable under the traditional tort law. Moreover, tort law is enforced by the courts, thus making the compensation unpredictable and unlimited. Radetzki (1999) argues, "traditional tort liability compensation pre-supposes not only that a certain act caused harm, but also that the harm is due to fault". These rules can be applied to nuclear liability as well. However, as discussed above, the nuclear technology development had been unique, and represents the efforts of a large number of entities and countries. Here pinning fault to any single person is almost impossible (IAEA 1997). As an example of the multinational nature of nuclear industry Raju and Ramana (2010) examined the status of Westinghouse as "Westinghouse was purchased by British Nuclear Fuels in 1998, which then sold it to Japan's Toshiba in 2006. In 2007, Toshiba sold 10 % of the company to Kazatomprom, the national uranium company for the Republic of Kazakhstan. Besides these, the Shaw group, based in Louisiana, US, owns 20 % of the company and the Japanese IHI Corporation a 3 % share". Owing to such an industry structure, in the nuclear accident scenario, the victims would find it difficult to establish fault of an entity, which could then be held liable. In case fault is established, the victims may have to ensure jurisdictional competence of the national courts (Schwartz 2006). It was thought desirable that a special regime was necessary—large and prompt compensation for the victims without any technicalities of tort, and full protection for industry through liability backed by insurance and governments. This initiative led to the development of the unique principles of nuclear liability law. Channelling the liability exclusively to the operator; absolute liability of the operator; exclusive jurisdiction to one national court of the Installation State; financial certainty of the liability, and liability limited in time and amount. Through this way, uniformity and harmonisation was sought and imposed.

Nuclear liability regime embarked upon a fundamentally different route on questions of the State liability for acts not prohibited by international law. On the question of international tort law for transboundary environmental damage, Sachs (2008) says, despite many decades of attempts, the international community could not facilitate tort remedies for victims of transboundary environmental damage. She concludes, though many civil liability treaties were concluded, only a few have entered into force, others continue to be "un-adopted orphans in international environmental law". The work by International Law Commission on 'International liability for injurious consequences arising out of acts not prohibited by international law' reveals the difficulty faced in attempting to codify laws on State liability (Brownlie 2008). Boyle (2005), in the context of sectoral agreements, states that nuclear energy liability conventions represent an alternative approach to transboundary liability. Countries enacted national legislation based on these unique principles, but with varying degrees based on national interests.

2.4.1 The United States Domestic Law

The 1957 Brookhaven report and the Enrico plant report—USAEC APDA-120 led to a popular belief that the development of nuclear power programs would be financially unviable. The Brookhaven report projected a worst case scenario, which assumed that 100 % of the core went airborne and drifted over Manhattan (Loewen 2011). In order to convince both the public and the industry, in 1957, the US Government through the enactment of the *Price-Anderson Nuclear Industries Indemnity Act 1957* (Price-Anderson Act) provided a legislative backing to nuclear industry to go ahead with the plans of power development. The Price-Anderson Act capped the cost of damages of a nuclear accident.

In drafting the indemnity plan, the US Congress through the original Price Anderson Act, 1974 established a two-tier insurance system. The primary layer required the operator to obtain insurance coverage of $60 million, with the secondary layer was provided by the government—amounting to $500 million. It has been argued that the US nuclear industry greatly benefited from having such a large provision of public fund for compensation (Faure and Vanden Borre 2008). Meehan (2012) quoting from hearings before the Senate Committee on Energy and Natural Resources, [107th Cong. 53, 54 (2001)] summarises, "the Act is essentially an insurance program that encourages private development of nuclear power, establishes a legal framework for handling potential liability claims, and provides a ready source of funds to compensate injured victims of nuclear accidents". This initial scheme witnessed two amendments in 1982 and 2005. With these amendments, a new scheme, i.e. retroactive premiums (explained below) completely replaced government indemnification with much higher compensation amount. From the two-tier scheme now a three-tier model is in place, the addition being industry-wide pool mechanisms requiring nuclear reactors to collectively contribute to a separate insurance pool.

Thus, under the current system, as the first tier, nuclear power plant owners pay premium for obtaining $375 million as insurance amount annually to private insurers, for each reactor unit. This is supplemented by a second tier retroactive premiums scheme. The second tier envisages, that if the primary layer is exhausted in the event a nuclear accident, "each licensee is required to provide a share up to $111.9 million" per reactor to cover costs in retrospective annual premiums (US NRC Fact sheet 2012). With 104 reactors currently licensed to operate, it is calculated that the second tier funds now contains about $11.6 billion (US NRC Fact sheet US 2012). When the second layer of compensation is also inadequate, the third tier kicks in—additional relief as determined by the US Congress. Such an elaborate public and private scheme is the uniqueness of US liability system. However, the Act has also been criticised by many. Green (1973) puts it very effectively that,

> The Price-Anderson is a technical measure necessary for adequate protection of the public interest with respect to a technology that exists and will inevitably grow substantially. The fact that the technology exists and grows only because of Price Anderson has been artfully concealed from public view so that consideration of the indemnity legislation would not trigger public debate as to whether nuclear power was needed and whether its risks were acceptable

Nevertheless, the uniqueness of Price Anderson Act lies in two of its mechanisms. The first being the concept of economic channelling and other is the industry insurance pool. US from the very beginning has followed a path of economic channelling of the liability, meaning, "that any entity may be held legally liable, but the economic consequences of that liability are channelled to the liable nuclear operator. Thus, any person who is held legally liable will be indemnified in respect of that liability" (Schwartz 2006). In essence, the US government has provided economic protection to all the supply contractors of NPPs, by fully exempting them from compensation pay-outs in cases of accident. The American Nuclear Society (2005) considers this a positive development which "by providing omnibus coverage, those who may be harmed are assured of the availability of funds to pay their claims, and firms that contribute in some manner to the design, construction, operation or maintenance of covered licensees are all protected".

The second important aspect of the Act is the industry insurance pool. Interestingly and unseen on other jurisdictions is the retroactive premiums scheme. Using this mechanism, the US government has effectively curtailed its financial obligation for paying compensation. The nuclear industry itself has been made responsible for arranging a large sum for making the payment at any given time. At hand for disposal is almost a $12 billion contributed by combined reactors owners operating within US. Thus, through this mechanism there now exists a substantial insurance amount funded by both the insurance and nuclear power industry. This could be a model for others countries. However, in the South Asian context, such a model may not have much relevance as all the NPPs are owned, operated and managed by government or government companies only.

In the initial years, the Price Anderson law had to go through several constitutional, legal and procedural challenges. The constitutional validity of the Act was challenged twice in, *Conservation Society of Southern Vermont v. Atomic Energy Commission* (No. 19-72 D.D.C. Apr. 16, 1975) and *Carolina Environmental Study Group v. Atomic Energy Commission* [438 U.S. 59. 1978]. In both these cases, the court upheld its constitutionality. However, the judgment had been contentious as well. Galiette (1978) was critical of the judgments; summaries his concerns as, "Nuclear power is representative of a new and qualitatively unique class of phenomena-radical technical innovation showing great promise but posing risks which may not be susceptible of reasonable assessment at the time of initial mass application." The law was first applied subsequent to the Three Mile Island accident in 1974 where compensation was paid.

Almost in parallel and taking cue from the enactment of the Price Anderson Act, several countries sought to legislate, domestically, on the twin issues of nuclear energy, i.e. development of the power program and addressing the liability concerns from a nuclear accident. In Europe this effort was led by OECD.

2.4.2 Organisation for European Economic Co-operation (OEEC)—Regional Efforts

In the aftermath of the Second World War, the Organisation for European Economic Co-operation (OEEC) was reconstituted in 1961 as the Organisation for Economic

Co-operation and Development (OECD). A grouping of the western European countries and the European Atomic Energy Community (Euratom), a grouping of six European Economic Council members, thought energy from nuclear sources could be a possible way to meet massive energy requirements in its re-construction efforts. The European Nuclear Energy Agency (ENEA) was established by a decision of the Council on 17 December 1957, and came into force on 1 February 1958. The ENEA was designed to meet three primary objectives, they are: (1) the establishment of joint research projects; (2) the solution of specific legal problems related to nuclear energy through the harmonisation of national laws or the adoption of regional conventions and (3) the provision of a forum in which the national nuclear energy programs of its member countries could be discussed and coordinated. With the membership of Japan, ENEA's name was changed in 1972 to the Nuclear Energy Agency (NEA), to reflect its growing membership beyond Europe's boundaries.

One amongst the first tasks under the OEEC was addressing a harmonised third-party liability for nuclear activities (Marcus 2008). Interestingly, even before the establishment of the specialised ENEA, the OEEC anticipated the issue of civil liability from nuclear energy accident and established committees in 1957 for nuclear liability and radiation protection (OECD-NEA 2008). The liability committee in the subsequent years evolved into the Group of Governmental Experts on Third Party Liability in the Field of Nuclear Energy[1]. This group became instrumental in drafting the *Paris Convention on Third Party Liability in the Field of Nuclear Energy* (Paris Convention) which was adopted by the OEEC Council on July 29, 1960, and it covers most of the West-European countries (OECD-NEA 2008). The Paris Convention entered into force on 1 April 1968 with 16 Contracting Parties. The Paris Convention thus became the first international instrument that established the rules of distribution of responsibility between countries, and ensuring compensation for damage caused by nuclear activities.

After the 1960 Paris Convention was adopted, it was recognised that the compensation amount that was provided for was too low. It is argued, "while governments at the time may not have envisaged a 'Chernobyl type' accident, they were very much aware that in the case of a nuclear catastrophe involving a large-scale emission of ionising radiation, hundreds or even thousands of people could suffer radiation-related illness, incur damage to their property and suffer various other forms of economic loss" (Schwartz 2006). In order to increase the financial coverage of the Paris Convention, the OECD countries created an additional finance mechanism to ensure that adequate compensation would be available to the victims of a nuclear accident. *The Convention Supplementary to the Paris Convention of 29 July 1960* (Brussels Convention) adopted in 1963 made the provision for additional or supplementary compensation for nuclear damage made up of 'public funds' in addition to the Paris Convention.

[1] A third party is anyone other than the nuclear operator itself and other than a supplier of goods, services or technology for use in connection with a nuclear installation. A third party may be inside or outside the nuclear installation and as such the term includes employees of the operator of the nuclear installation at which an accident occurs (Schwartz 2006).

In the following years, both the Paris Convention and the Brussels Conventions have been further strengthened thrice: by Additional Protocols adopted in 1964, 1982 and 2004. Contracting Parties to both the Paris and the Brussels Conventions have enacted respective domestic legislations which implemented the provisions of the 1960 Paris and the 1963 Brussels Convention.

2.4.3 IAEA Sponsored International Framework

With the adoption of the Paris Convention, internationally, a need was felt for a global liability regime with the aim of harmonising national legislations pertaining to third-party liability for nuclear damage. This was particularly important for two reasons. Firstly, countries were entering into bilateral agreements in faraway places without clear understanding of the rules; secondly, applicability of the Paris Convention was limited to West-European countries.

The international regime was multifaceted in respect to jurisdiction and choice of laws that an accident in one State, may result in claims and suits in many other States, which may end up without providing for compensation (IAEA 1960). The IAEA was convinced on a Convention that "can serve as a basis for harmonized and largely uniform rules regarding civil liability for nuclear hazards. Such a convention should bind not only the States in which nuclear energy is now utilized but also others in which damage may be suffered or where nuclear industry is expected to develop in the future" (IAEA 1960).

In April–May 1963, an international conference approved the *Vienna Convention on Civil Liability for Nuclear Damage*, (Vienna Convention), which was opened for signature on 21 May 1963, and came into force on 12 November 1977. It took 15 years for the Vienna Convention to come into force even though only five states were required to ratify it. The delay has been mainly due to the perceived commonality with the Paris Convention. Major nuclear States had already been committed to the process of the Paris Convention. A detailed analysis of the Vienna Convention is made in the later part of this chapter.

With the adoption of the Vienna Convention, two parallel conventions existed, neither applied to nuclear damage suffered in the territory of a party to the other. Only in the aftermath of the Chernobyl nuclear disaster in 1986, did the world community take concrete steps to reconcile the differences between the Paris and Vienna Conventions and also to strengthen the foundation of nuclear liability principles.

2.5 The Chernobyl Nuclear Disaster and State Liability

A catastrophic consequence from a nuclear power accident came true on 26 April 1986 at Chernobyl, in the former USSR. The radiation exposure caused serious social and economic disruption, covering large populations in Belarus, Russia and Ukraine (UNSEAR 2008). The world did not know for many months the exact chain of events leading to this disaster (Malone 1987). Literatures point out that the radioactive plume covered much of Europe; even reaching the United Kingdom in the first few days of May 1986 (Woodliffe 1990).

At the time of the Chernobyl incident, two international civil nuclear liability conventions were in force—the 1960 Paris Convention and the 1963 Vienna Convention. The USSR was not a party to either of these Conventions. Unfortunately, the USSR did not notify its neighbours on the occurrence of the accident as well. Chernobyl was classified as Level 7[12] event on the International Nuclear Event Scale by the IAEA—the highest possible classification. The accident fully brought home the picture of the transnational consequences of a nuclear accident—that the geographical scope is not necessarily confined to national boundaries.

Subsequent to the accident, the countries sought in various ways to hold the USSR liable for compensation. At the time of Chernobyl there was no international treaty relating to State liability. However, customary principles of State Responsibility had evolved that could make nations accountable for their actions and two early cases represented this are the *Trail Smelter Arbitration* (1939)[3] and *Corfu Chanel Case* (1949).[4]

Wood (2004) states that the Trail Smelter Arbitration is "remembered as the earliest articulation of two core principles of international environmental law: that states have a duty to prevent transboundary environmental harm, and that they have an obligation to pay compensation for the harm they cause". The Arbitration awarded monetary damages to US for the actions of Canada. In the *Corfu Channel* case, the International Court of Justice (ICJ) found that Albania had an obligation to notify the approaching ships of the dangerous condition on account of mine fields in the area. The court held the Albanian government responsible under international law for the explosions, and for the resulting loss of life and damages (ICJ

---

[2] International Nuclear Event Scale (INES) is a tool informing the magnitude of a disaster. The accidents are classified from level '0'–'7'. Level '0' has no safety significance. Level '7' is classified as a major accident. Chernobyl and Fukushima accidents are of level 7. Other levels of classification are, '1'-anomaly; '2'-incident; '3'-serious incident; '4'-accident with local consequences; '5'-accident with wider consequences; '6'-serious accident. Basically, levels 1–3 are incidents and levels 4–7 are accidents (IAEA 2012).

[3] The US alleged that the fumes from the operation of Canadian Smelter Company caused pollution (nuisance) to Washington State. The tribunal that was set up to try the case concluded that Canada was responsible for transboundary pollution and owed damages to US.

[4] Two British destroyers struck mines in Corfu Strait in Albanian waters and suffered damage, including serious loss of life. UK filed an application against Albania in ICJ seeking the Albanian Government to take responsibility and also pay compensation. Albania contented that the UK had violated Albanian territorial waters. ICJ found that Albania was responsible for the explosions and for the resulting damage and loss of human life suffered by the United Kingdom. The Court also found merits with Albania that minesweeping by UK had violated its sovereignty. ICJ ordered Albania to pay the United Kingdom.

1949). Further, in respect to damages to the environment, in 1972 the UN through General Assembly Resolution 2996, in its plenary meeting on 15 December 1972 re-emphasised the importance of Principle 21 ground rules which govern the responsibility of the States to the international arena in preserving and protecting the environment (UN 1972).

Taken altogether, there was a case for the neighbouring States to seek legal remedies from the USSR. However, the USSR had not consented to the jurisdiction of the ICJ at that time and moreover, the decisions of the ICJ are enforced by the Security Council where the USSR has a veto power (Malone 1987). Further, the USSR was not a party to any of the liability conventions. On account of all these factors, none of the affected States brought an action based on public international law for recovery (Malone 1987; Heiss 1993). One may remember that even for the famed Trail Smelter Arbitration award, on account of various factors, it took 13 years before the victims were paid compensatory damages for the injuries suffered (Kaplan 1988).

The legal consequences of the Chernobyl accident were immense. Here was a case, wherein a nuclear accident had adversely affected a large number of people in a geographic area that transcend boundaries, without any legal remedy available. It became apparent that any chance of establishing a worldwide applicable liability law would require the cooperation of all States. Cooperation would require ratification of international treaties and harmonisation of laws with international principles. In the face of Chernobyl; the Paris and Vienna Conventions were seen as ineffective systems requiring considerable reforms for worldwide acceptability.

Nuclear reactor technology safety features and liability and compensation were subject to intense international scrutiny. ElBaradei (2007) stated, "Chernobyl, imprinted on public consciousness as proof that nuclear safety was an oxymoron". In order to overcome legal lacunas, significant changes were mooted to both the 1963 Vienna Convention and the 1960 Paris Regional Convention liability regimes. Moreover, to address the overlap due to the existence of two parallel conventions, a new Protocol *1988 Joint Protocol Relating to the Application of the Vienna Convention and the Paris Convention* (Joint Protocol) was established to work as a link between these two conventions. Also, a new international liability regime, *Convention on Supplementary Compensation for Nuclear Damage* (Compensation Convention or CSC) was also adopted. In this reforms initiative, while strengthening the regime, the major focus was to ensure universal adherence to liability law.

Amongst others, the lesson learned from Chernobyl was that if information on the accident was provided on time and also assistance was given early, lives and livelihood could be saved. Under the aegis of the IAEA, two conventions were drawn up and quickly adopted in 1986,—the *Convention on Early Notification of a Nuclear Accident* (Early Notification Convention) and *the Convention on Assistance in the Case of a Nuclear Accident or Radiological Emergency* (Assistance Convention).

2.6 Chernobyl Effect: Reforms in International Nuclear Liability Law

2.6.1 Emergency Conventions

The two international conventions drawn up with remarkable speed under the auspices of the IAEA are: (1) the Notification Convention and (2) the Assistance Convention. Both these Conventions together referred to as 'Emergency Conventions' were adopted on the same day, 26 September 1986.

The Early Notification Convention requires the States to notify nuclear accidents which have the potential for international transboundary release that could be of radiological safety significance for another State. Reporting of a nuclear accident is mandatory, involving facilities and activities listed in Article 1. The Assistance Convention on the other hand, sets out an international framework for co-operation among the State Parties and with the IAEA, to facilitate prompt assistance and support in the event of nuclear accidents or radiological emergencies. It requires the States to notify the IAEA of their available experts, equipments, and other materials for providing assistance. In case of a request, each State Party decides whether it can render the requested assistance as well as its scope and terms. Assistance may be offered without cost, taking into account *inter alia* the needs of developing countries and the particular needs of countries without nuclear facilities. The IAEA serves as the focal point for such cooperation by channelling information, supporting efforts and providing its available services. Liability and compensation are not within the purview of the Emergency Conventions.

2.6.2 Joint Protocol Linking the Paris and the Vienna Liability Conventions

One of the major complexities in the international liability regime in the pre-Chernobyl era was the relation between the Paris Convention and the Vienna Conventions; countries had a choice of treaty adherence. Both these treaties did not recognise each other's jurisdiction, which raised many questions. The IAEA in the explanatory text, states that two areas were crucial for the effectiveness of international nuclear liability regime. First, "importance of having a universally acceptable regime; and second, the adequacy of the regime to cope with the trans-boundary consequences of a major nuclear accident" (IAEA 2004). On account of this, linking these two conventions became necessary.

This motivated both the Paris and the Vienna Convention States to establish a formal relationship between the two instruments. Expert groups established by the IAEA in cooperation with the OECD Nuclear Energy Agency (OECD NEA), adopted the Joint Protocol with the intent to link the two conventions. The Joint Protocol entered into force on 27 April 1992.

Schwartz (2006) summarises the primary aim of the Joint Protocol, as being to "remove conflicts resulting from the simultaneous application of both Conventions while promoting broader adherence to the basic principles underlying those Conventions". As of date, 27 States are parties to the convention.

Interestingly, the purpose of the Joint Protocol has not been fully met, since many of the States having nuclear power programs continue to adhere only to either of the Conventions. It is noteworthy, that even today Russia and many other countries in Europe including the UK and France—offending and the victim States of the Chernobyl accident, have not signed the Joint Protocol. This means that even under present circumstances, countries could refuse State liability and payment of compensation on account of being non-party to the Joint Protocol. Moreover, soon after the adoption of the Joint Protocol, it was realised that this protocol is not adequate to cover compensation from accidents such as Chernobyl. Joint Protocol only provided "a solution regarding the relationship between the victims in Vienna and Paris States" (Pelzer 2010). Significant increase in the amount of compensation and also issues of States that are not party to any of the conventions, remain as issues not addressed by the Joint Protocol. In order to make the regime more broad based, comprehensive reforms are necessary in the international nuclear liability system.

2.6.3 Amendments to the Vienna and the Paris Conventions, and the CSC

The 1986 Chernobyl accident prompted widespread awareness of the need for improved protection of the public from the consequences of nuclear accidents (Boulanenkov and Brands 1988). It demonstrated that both the Paris and the Vienna Conventions' provisions were inadequate in many ways. There was a need to increase the amounts of liability, broaden the types of damage, definition of damage, jurisdiction, equal protection of victims in any country etc.; all requiring substantial changes in the existing regimes. In short, the current legal regime requires modernization (Lamm 1998). The rallying point was that, a nuclear accident having transboundary impact would require equal protection of victims wherever they are—contracting and non-contracting parties. Moreover, the existence of multiple regimes undervalues the importance of the regimes themselves. Significant changes were proposed for both the Paris and the Vienna Conventions.

Internationally, the IAEA "followed a two-track approach: to improve the existing civil liability regime, including revision of the Vienna Convention for which the IAEA is depositary; and, to develop a comprehensive international liability regime" (IAEA Explanatory Text 1997). After many years of negotiations, under the aegis of the IAEA, in 1997 delegates from over 80 States adopted a *Protocol to Amend the 1963 Vienna Convention on Civil Liability for Nuclear Damage* (1997 Protocol) and also Convention on Supplementary Compensation, known as CSC. Regionally, the Paris Convention countries took close to 10 years to reflect the much required changes. On 12 February 2004, *the Protocol to Amend the Paris Convention on Third Party Liability in the Field of Nuclear Energy of 29 July 1960* (2004 Protocol) was adopted. On account of the changes in the Paris Convention, the Brussels Convention was also amended through, *the Protocol to Amend the Brussels Convention Supplementary to the Paris Convention of 29 July 1960* (2004 Brussels Protocol). To an extent, the revisions in the Paris and the Brussels Conventions are aligned, and made them compatible with the 1997 Vienna Protocol and CSC (Schwartz 2010).

In respect to extent of liability amount, both the protocols have significantly increased liability amounts. The 1997 Vienna Protocol increased operator's liability to not less than 300 million Special Drawing Rights (SDRs), which is approximately 400 million US dollars (Article V). In the case of CSC as well, States can provide for the maximum or a 'greater amount' [Article 3(a)(1)]. This flexibility was provided to accommodate many States that did not wish to have a limited liability regime. In case unlimited liability is included, the financial security requirements will be capped at 300 million SDR for the operator (Article VII 1997 Protocol and Article III CSC). CSC provides a second tier compensation from the pooled funds of the contracting parties (Article III(1)(b). In the case of the 2004 Paris Protocol, the amount of a nuclear operator's liability has been significantly increased to an amount not less than Euro 700 million (US$ 935.45 million) (Article VII). Supplementing this, the revised 2004 Brussels Protocol in the second tier raised to Euro 500 million by the installation State and the third tier from the funds pooled by the contracting States increased to Euro 500 million—combined total of Euro 1.5 billion (Article 3). Interestingly to accommodate unlimited liability States like Germany, Austria and Switzerland; the 2004 Protocol recognises that a State with an unlimited liability regime may equally participate in the scheme established by the Convention.

Thus, under the revised protocols, major increase in amounts as well as attempts for harmonious existence of both limited liability and unlimited liability regime were made. However, the increased minimum amount and the classification of limited and unlimited liabilities in the same Conventions have now contributed to the new fear of disparity in liability amount (Schwartz 2010). The States having unlimited liability regime on the other hand continue to argue that the provision for limited liability must end, since limited liability is seen as a subsidy given to nuclear industry (Currie 2008). Researchers have argued that there exist good reasons for providing such subsidy during the initial phase of a nuclear power program, but not when the industry has matured. Thus, countries are in a way demanding reforms in the founding principles as well.

One of the major successes of the 1997 Protocol was its geographical coverage. Nuclear damage 'wherever' they occur will be within the scope of the Protocol [Article 1(A)(1)]. The Installation State can exclude the applicability of the Protocol to a non-contracting State with nuclear installation that does not provide reciprocal benefits [Article 1(A)(2 & 3)]. The 2004 Paris Protocol also witnessed an expanded scope of geographical coverage. Largely similar to the 1997 Protocol, but the difference here is in its applicability based on Convention State and reciprocity based on the Paris Convention Principles [Article 2(a)]. The 2004 protocol text states that the applicability in a non-Convention State is limited if such a State, (a) is a party to the Vienna Convention and the 1988 Joint Protocol, or (b) it has no nuclear installations or (c) its nuclear liability legislation affords equivalent reciprocal benefits and is based on principles identical to those contained in the Paris Convention.

This provision of the 'exclusion' of States, whatever may be the classification in both the revised regimes can be counted as one of the reasons that the 1997 Protocol has still not seen major adherence. It is interesting to note that during the negotiations for the 1997 protocol, many States argued for a blanket geographical scope (IAEA 1997). The compromise language was the scope 'wherever' subjected to domestic application of law. In respect of the 2004 Protocol, considering Europe, the major coverage of the 2004 Protocol are the States that do not come under the expanded geographical scope, i.e. party to the Vienna Convention and the Joint Protocol.

The 1997 Vienna Protocol [Article I(1)(f)], Compensation Convention (Article I(f)) and the 2004 Paris Protocol (Article 1(a)(vii)) have broadly provided for alike definition of nuclear damage. The 2004 Paris Protocol does not have the definition provided under (viii). Though, in the aftermath of the Chernobyl accident, adopting such an expanded definition was seen as a requirement, but the negotiations had to endure serious disagreements by the States which wanted the definition to be aligned with their respective national laws (Soljan 2000 quoting Reyners 1998). Consensus prevailed on the need for a comprehensive definition, and the protocol now defines nuclear damage as:

(k) "Nuclear Damage" means:

(i) loss of life or personal injury;
(ii) loss of or damage to property;

and each of the following to the extent determined by the law of the competent court:

(iii) economic loss arising from loss or damage referred to in para (i) or (ii), insofar as not included in those sub-paragraphs, if incurred by a person entitled to claim in respect of such loss or damage;
(iv) the costs of measures of reinstatement of impaired environment, unless such impairment is insignificant, if such measures are actually taken or to be taken, and insofar as not included in para (ii);
(v) loss of income deriving from an economic interest in any use or enjoyment of the environment, incurred as a result of a significant impairment of that environment, and insofar as not included in para (ii);
(vi) the costs of preventive measures, and further loss or damage caused by such measures;
(vii) any other economic loss, other than any caused by the impairment of the environment, if permitted by the general law on civil liability of the competent court.

Lamm (2006) in the context of the 1997 Protocol views "redefinition by the 1997 Protocol is clearly reflective of an intention to ensure as possible full compensation to victims of nuclear damage". Though well intended, such an expanded definition both in the 1997 Protocol and the Compensation Convention became problematic due to the wording, "the extent determined by the law of the competent court". In effect, allowing all types of damages to be included in the definition. Soljan (2000) states that "The formulation of this definition is not entirely clear. The reference to the 'extent determined by the competent court' does not lead to the conclusion that the law of the competent court is applicable to the question of the admissibility of the claim for compensation for economic loss, environmental damage or damage to the property". This can be viewed as expecting too much from the national courts of the Installation State, to interpret what constitutes these damages and how will the compensation be distributed. Further, national standards relating to environmental pollution and their interpretation by the national courts are never uniform. For example, today, India can boast of multiple legislations covering environmental issues, but none deals with civil liability and compensation. Courts have been interpreting the liability through very liberal tort principles. Owing to such a situation, the application of this protocol definition in the South Asian context, as well as other regions of the world will be complex. The result would lead to unpredictable transboundary claims.

The 2004 Paris Protocol does not include 'economic loss' or 'the extent determined by the law of the competent court' in its definition. The exclusion of economic loss, as argued by Schwartz (2010), has resulted in incompatibility and non-applicability in relation to the 1988 Joint Protocol and the Compensation Convention—in effect does not obligate the Paris Convention States to compensate even when the victims belong to a Paris Convention State.

Another interesting inclusion in the 1997 Protocol is the 'priority principle'. The principle has not received adequate critical attention from researchers. Article 10 of the Protocol states, "where in respect of claims brought against the operator the damage to be compensated under this Convention exceeds, or is likely to exceed, the maximum amount made available pursuant to paragraph 1 of Article V, priority in the distribution of the compensation shall be given to claims in respect of loss of life or personal injury". The Compensation Convention on the other hand in Article III(2) specifies non-discrimination. It provides that "compensation shall be distributed equitably without discrimination on the basis of nationality, domicile or residence". This is qualified by the provision that the law of the installation State may exclude non-contracting States [Article III(2)(a)]. In effect, since the CSC may not be the preferred law in South Asia, applicability of this non-discrimination principle is also in doubt.

The 2004 Protocol on the other hand does not refer to priority principle. But a review of the domestic laws of the Paris Convention States provides fine details of application of the priority principle. Spain, France, Poland, Norway, Netherlands, Slovenia, Bulgaria provide for the same, while the UK, Czech Republic, Denmark, Norway, Hungary, Italy do not contain any priority principles (OECD-NEA 2009).

In the case of Germany, the law provides an ordinance that establishes criteria for the distribution of compensation (OECD-NEA 2009).

The priority principle basically entails that in cases where the total amount of compensation payable is more than the amount available under the convention; priority will be given to claims relating to loss of life and personal injury. An argument can be raised that, this favours the Installation State. The reason being that in case of an accident, the major impact will be borne by the area immediate to the NPP, i.e. within the Installation State itself. In case of existence of a trans-boundary impact, the consequences will be largely on land, water etc.—essentially economic loss. In case of loss of life and personal injury, the most-affected State will receive a priority in obtaining compensation, while others may have to wait or may not get it at all. The additional question is how the courts of the Installation State will decide the quantum of compensation on the priority. Warren (2000) observes, "the question of how the compensation fund is to be administered in an equitable and prioritised manner and by whom and at what cost of settlement is one which I consider to be one of the most glaring oversights in the international conventions on nuclear liability". Though the inclusion of the priority principle provision is seen as a major reform in identifying the priority areas for compensation, it ended up as a discriminatory disbursement regime of compensation between the States.

The 1997 and 2004 protocols may have strengthened both the Vienna and the Paris Conventions to an extent, but also in equal measure created problems in fulfilling the harmonisation objectives. For example, a State can become party to the 1997 Protocol without ratifying the 1963 Vienna Convention. Such a Contracting Party will be bound by the provisions of the 1963 Convention, in relation to other State Parties to the Protocol. Currie (2008) views the situation as, "in other words, Parties which join the Protocol but not the Convention are bound by the lower limits in the Vienna Convention unless they state otherwise at the outset, but Parties which join the Convention but not the Protocol are not bound by the higher limits of the Protocol in any event".

On the question of compatibility, Adisianya (2011) says, these new conventions (referring to both the 1997 Protocol and the 2004 Paris Convention Protocol) "have, in effect, created and broadened the problems which the Joint Protocol intended to solve the situation of no complete link amongst the conventions is more complex and complicated". A country now can join the 1997 Protocol without being a party to the 1963 Convention, thereby creating confusion within the Joint Protocol signatories. In respect to adherence, sadly only 12 countries are currently party to the 1997 Protocol: Argentina, Belarus, Bosnia and Herzegovina, Jordan, Latvia, Morocco, Romania, United Arab Emirates, Saudi Arabia, Poland, Montenegro and Kazakhstan. None of these countries have significant nuclear power programs. Russia which had signed the treaty in 1996 ratified it only in 2005. Ukraine, the country most affected by the Chernobyl accident has still not ratified the Protocol. In the case of the 2004 Paris Protocol (not in force), the situation is grimmer. Except Norway and Switzerland, none of the other Paris Convention states have ratified the 2004 Protocol.

2.7 Continuing Efforts in Liability Regime Reforms

In view of the non-adherence by the States, especially in the European Union (EU), the official position calls for greater harmonisation through involvement of the EU. The communication from the European Commission (EC) to the Council and the European Parliament—Nuclear Illustrative Programme (NIP) presented under Article 40 of the Euratom Treaty describes the difficulties within EU (European Commission 2007). On account of the eastern expansion and as a majority of the new States of the EU follow the Vienna Convention, the EC observes, "The Commission is aiming at harmonising the nuclear liability rules within the Community…. In order to finalise and improve the proposals already made, the discussion should notably focus on developing a harmonised liability scheme and mechanism to ensure the availability of funds in the event of damage caused by a nuclear accident". The European Economic and Social Committees in their opinion to NIP (EESC 12 July 2007) say, "for greater acceptability of nuclear power, the current system (liability insurance of EUR 700 million) is inadequate for this purpose". In fact some of the recent literatures from Europe (Handrlica 2010) advocate further EU involvement in the area of nuclear liability by means of a European Nuclear Liability Directive.

In Europe, there are already strong signals of reform. Austria in 1998 enacted a new Federal Law on *Third Party Liability for Nuclear Damages caused by Radioactivity* where in the founding principles were completely transformed. Hinteregger (1998) summarises the changes as, "liability is unlimited in amount, legal channelling is to a great extent eliminated, there is no exclusive jurisdiction—Austrian court has jurisdiction and Austrian law will be applicable if nuclear damage occurs in Austria, regardless of where it was caused". The Austrian law should be seen in the context of countries tired of waiting for constructive reforms. In the developing world, India has enacted its *Civil Liability for Nuclear Damage Act 2010* (discussed in detail in Chap. 3).

In 2011, UK announced its intent to harmonise the *Nuclear Installations Act 1965* in tune with the revised 2004 Paris and Brussels Protocols (UK Department of Energy and Climate Change ("DECC") 2011). UK government has also made it clear that they will not apply the priority principles. One may note that such reforms initiatives are an exception. The main changes reflected in the revised law would include (DECC 2012):

> Increase in the categories of damage for which operators are liable include, "personal injury and loss of life ("personal injury") and property damage (i) economic loss arising from property damage; (ii) the costs of measures of reinstatement of the impaired

environment; (iii) loss of income deriving from a direct economic interest in any use or enjoyment of the environment; and (iv) the costs of preventive measures".
Broadening the geographical scope to damage suffered in certain non-Paris countries (non-nuclear countries and certain other countries that have equivalent reciprocal arrangements) as well as the UK and other Paris countries.
A significant increase in the financial liability of the operator from currently 140 million pounds to Euro 1200 million.

As expected, it took another nuclear disaster to restart the comatose negotiations towards achieving a universal and harmonised liability regime.

2.7.1 Fukushima Nuclear Disaster

On 11 March 2011, a 9.0 Richter scale magnitude earthquake hit northeastern Japan resulting in a tsunami, with waves as high as 30–38 m (IAEA 2011). The earthquake and tsunami combined caused the blackout of four Nuclear Power Plants at Fukushima Daiichi (Fukushima accident). OCED Fukushima Press kit (2012) states that at the plant site, the tsunami was estimated to be 14–15 m; the plant had been designed to withstand upto 4–5 m. The accident was rated as Level 7 on the International Nuclear Event Scale, the same classification as of Chernobyl. On account of explosions in the plants, reports suggest there was radiation release to the environment. People living around the plant up to 20 km were compulsorily evacuated by the government (Prime Minister Office Press Briefing 2011). A "planned evacuation zone" of 30–50 km zone, north of the plant was also established (Sugimoto et al. 2012).

Osaka (2012) states that major Japanese industries such as—agriculture, fishing, tourism, and other businesses were heavily impacted. The US had put import restrictions of sand lance (a small fish) from Fukushima Prefecture (Buck and Upton 2012). Since the accident occurred on the eastern side of Japan, bordering the Pacific Ocean there has been no substantial transboundary impact on other countries. Few reports suggest considerable amount of radiation has entered into sea, however, none of the reports confirm that the impact could be disastrous (Buck and Upton 2012; and Madigana et al. 2012). A UNSCEAR (2012) press release, suggests that it may take years to know the full impact. As such is the case of transboundary damage, so far, liability and compensation issues have been limited only to Japanese jurisdiction (OCED Fukushima Kit 2011). It is interesting to note that Japan has not acceded to any of the international nuclear liability conventions when the accident occurred. However, on 15 January 2015 Japan joined CSC and with its acceptance, the CSC will be effective from 15 April 2015. They have their own domestic legislation, without reciprocity. Had the accident occurred in the west of the country, though highly unlikely, considerable damage could have been caused in Korea. The Fukushima nuclear accident emphasises the need for undertaking reforms that are acceptable to more countries.

The current slow pace of reforms on the liability regime, reflects the inability of the international community to achieve a universal harmonised regime. Based on this, Pelzer (2010) argues "even if one accepts that a globally unified liability regime most probably is not achievable and that adherence to the international regime by a number of States is 'not even wanted' there is nevertheless the urgent need to strive for harmonisation in larger geographical regions which qualify for establishing risk community". He views that the recent developments from the EC; the preparations for establishing a unified nuclear liability regime within the EU member States as most desirable. This is the route South Asia should consider—formation of a 'South Asian Nuclear Risk Community'.

Cooper (2011) sums-up the state of affairs, "The need to revisit and revise regulations regarding financial responsibility for nuclear accidents has been clear and compelling for at least a quarter of a century (since Chernobyl) and has been made overwhelmingly obvious by Fukushima". Also Reyners (2013) hopes that the Fukushima accident could serve as a catalyst for real progress in the nuclear liability regime. The international community expects that the Fukushima accident will lead to real progress.

The Fukushima nuclear accident transformed the discussion on nuclear risk paradigm. The IAEA (July 2012 and August 2012) in response to the accident, re-emphasise the need to achieve a 'global nuclear liability regime'. To take stock of the safety of nuclear energy in the wake of Fukushima accident, the IAEA in June 2011 convened a Ministerial Conference and adopted a 'Draft Action Plan on Nuclear Safety' (Action Plan) (IAEA 2011). On the question of nuclear liability regime the IAEA Action Plan states,

> Member States to work towards establishing a global nuclear liability regime that addresses the concerns of all States that might be affected by a nuclear accident with a view to providing appropriate compensation for nuclear damage. The IAEA International Expert Group on Nuclear Liability (INLEX) to recommend actions to facilitate achievement of such a global regime. Member States to give due consideration to the possibility of joining the international nuclear liability instruments as a step towards achieving such a global regime.

Taking the initiative put forward by the IAEA Action Plan, the International Expert Group on Nuclear Liability (INLEX) recommended actions to achieve such a global regime (IAEA 2012). Overall, the recommendations request the States to participate in the existing international regimes, taking advantage of the higher level of flexibility offered by these conventions. A review of the recommendations suggest, either it was not given a comprehensive terms of reference or it did not thread the difficult part of reforms. Continuing to emphasis on adherence when countries have serious difficulties on becoming party of the present regime is a distressing sign from a world body.

In Europe, the European nuclear plant's 'stress test' emphasises the central role played by EC, particularly calling countries to legislate on liability and insurance. In the case of South Asia, a case is made in the subsequent chapters that South Asia being a nuclear-risk zone, the regional forum—the South Asian Association for Regional Cooperation (SAARC) in view of its standing should establish a forum to discuss nuclear safety and liability issues. The reason being, South Asia is largely a single geographical unit which is politically and culturally connected. Though

India remains the largest in size and a link to all other countries, the nuclear power program that is either planned or have ambitions by four countries—Bangladesh, India, Pakistan and Sri Lanka out of eight SAARC nations draws a parallel to the European nuclear dilemma—geographical proximity and actions of a responsible state. In the nuclear liability discourse, a major drawback is that there is little consideration of the voice of non-nuclear countries. In South Asia, Nepal, Bhutan and Maldives, are unlikely to initiate a nuclear program but could be adversely affected in case of a nuclear disaster. As the mapping exercise in Chap. 5 shows during an accident, these non-nuclear countries could well be part of the radiation exposure zone. A detailed discussion of the importance of SAARC, a regional forum within South Asia, though not a legal union, (unlike the EU) towards the establishment of a regional nuclear risk community is discussed in Chap. 4.

2.7.2 Nuclear Exporter's Principle: Achieving the Impossible?

Parallel to government and intergovernment efforts, the Carnegie Endowment for International Peace (CEIP) in 2011 led the world's major civilian nuclear power plant vendors (excluding India and China) in the development of a common understanding of the parameters that should guide the exporters of nuclear power plants (http://nuclearprinciples.org/about/history/). These principles are referred to as the 'Nuclear Power Plant Exporters' Principles of Conduct". The principle on Compensation for Nuclear Damage states:

> Before entering into a contract to supply a nuclear power plant to a Customer, the Vendor will independently make a reasonable judgment that the Customer State has in force, or will have in force before fuel is delivered in the Customer State's territory, a legal regime providing adequate and prompt compensation for the public in the unlikely event of an accident, with protection in effect equivalent to one or more of the following best practices:
>
> 4.1 A legal regime for compensation and nuclear liability that, *inter alia*:
>
> 4.1.1 Contains adequate liability limits and financial protection consistent with current international standards;
> 4.1.2 Is backed by Customer State guarantees;
> 4.1.3 Ensures that claims for compensation by possible victims will be channelled to the operator of the nuclear power plant(s) that would be strictly and exclusively liable and channelled to one single competent court;
> 4.1.4 Includes compensation for personal injury, property damage, environmental damage, loss of income, economic loss, and preventive measures;
> 4.1.5 Does not allow compensation amounts to be set aside or reduced by unilateral strict reciprocity requirements; and/or
>
> 4.2 A treaty relationship with the Vendor State under either the IAEA's Vienna Convention on Civil Liability for Nuclear Damage, as amended or, if eligible, the Organisation for Economic Cooperation and Development's Paris Convention on Third Party Liability in the Field of Nuclear Energy, as amended; and/or

4.3 The IAEA's Convention on Supplementary Compensation for Nuclear Damage (CSC)—which is the IAEA's unified global nuclear liability regime that any State can join if it is a Party to the Vienna Convention or Paris Convention or has a domestic law that is consistent with the CSC Annex. Such action would enable global treaty relations crucial to assure worldwide compensation and liability protection during plant operation and transnational transport

CEIP's exercise has been unprecedented, considering the nature of industry. All major exporters of nuclear power plants are party to these voluntary Principles. The principles have been adopted by nine companies based in Canada, France, Japan, Russia, South Korea and the United States—all having major stake in world exports, and to India in particular. These principles are the outcome of a voluntary initiative; are in the nature of a soft instrument and thus Principles, and are not legally binding. Even so, each company has independently undertaken to implement the Principles in the course of its business activities. In practise, this is easier said than done, as the business opportunities may occur in regions, with legal and political systems that are at variance. By way of example, India's liability law (*The Liability for Civil Nuclear Damage Act 2010*) is seen as being inconsistent with international liability regime. India on the other hand has argued that the law strengthens the international regime (Chap. 3 discusses in detail India's liability law). Further, there is a view from countries like India that these Principles are seen as a strategic move against companies from India, hindering them from exploiting lucrative export markets. Countries could strategically restrict imports if such companies are not part of 'Principles of Conduct' (Ram Mohan 2011). France which has a major business interest in India, has in some ways already broken away from the group. The President of France during an interview stated, "Regarding civil nuclear liability, we obviously respect Indian law. It is the sovereign decision of a country that has witnessed catastrophes like the Bhopal gas tragedy" (French Embassy in New Delhi 2013).

2.8 Conclusion

Many countries have embarked on the journey of harnessing energy from nuclear resources, even after witnessing the consequences of the bombings in Japan and the Chernobyl accident. Technology supplier countries argued that though studies deliberate the probable magnitude of an accident, such outcomes are rare. Interestingly, the supplier countries ring fenced themselves from possible liability claims through the use of bilateral treaty provisions, domestic legislations and also through advocating adoption of unique principles, in particular liability chanelling and limits on liability. Many countries have adopted these founding principles in their domestic legislations. The Chernobyl nuclear power plant accident was an eye-opener. States realised that the existing regime was entirely incapable of dealing with nuclear disasters of such magnitude. Major reforms were agreed to and brought about by establishing a new regime and strengthening old liability regimes.

However, today none of the reforms can be termed as successful. Adherences to these reforms are scantier than the original conventions. The 1997 Vienna Protocol has only 10 ratifications. The purpose of the Joint Protocol has been defeated; Europe which had faced the brunt of the Chernobyl disaster still does not have a harmonised liability regime, and many countries continue to adhere to only the original conventions. The 2004 Protocol to Paris Convention has only 2 contracting Parties (Norway and Switzerland).

Even after 25 years of Chernobyl, the international nuclear liability regime is extremely inconsistent and does not inspire worldwide confidence. The goal of harmonisation and universal adherence still has a long way to cover. Fukushima has given yet another chance to reform the regime.

References

Adisianya, A. (2011). Different Compensation Systems Under Nuclear Liability Conventions. *CEPMLP-Dundee*, vol. 14. Retrieved March 16, 2012, from www.dundee.ac.uk/cepmlp/gateway/files.php?file=cepmlp…76.

American Nuclear Society. (2005). *The Price-Anderson Act: Background Information*. Retrieved February 19, 2013, from http://www.ans.org/pi/ps/docs/ps54-bi.pdf.

Anderson, D. R. (1978). Limits on liability: The Price-Anderson Act versus other laws. *The Journal of Risk and Insurance, 45*(4), 651–674.

Bathurst, M. (1947). Legal aspects of the international control of atomic energy. *British Yearbook of International Law, 24*, 1–32 (Oxford University Press).

Bechhoefer, B. G., & Stein, E. (1957). Atoms for Peace: The new international atomic energy agency. *Michigan Law Review, 55*(6), 747–798.

Benoit-Smullyan, E. (1946). Control of atomic energy by the United Nations. *The Antioch Review, 6*(4), 488–494.

Boulanenkov, V., & Brands, B. (1988). Nuclear liability: Status and prospects. *IAEA Bulletin, 30*(4). Retrieved June 22, 2011, from http://www.iaea.org/Publications/Magazines/Bulletin/Bull304/30405380409.pdf.

Boyle, A. E. (2005). Globalising environmental liability: The interplay of national and international law. *Journal of Environmental Law, 17*(1), 3–26.

Brownlie, I. (2008). *Principles of public international law*. London: Oxford University Press.

Buck, E. H., & Upton, H. F. (2012). Effects of Tohoku Tsunami and Fukushima radiation on the U.S. marine environment. *US Congressional Research Service*. Retrieved September 05, 2013, from http://fas.org/sgp/crs/misc/R41751.pdf.

Buck, A. (1983). The Atomic Energy Commission. U.S. Department of Energy. Retrieved May 10, 2011, from http://energy.gov/sites/prod/files/AEC%20History.pdf.

Cavers, D. F. (1964). Improving financial protection of the public against the hazards of nuclear power. *Harvard Law Review, 77*(4), 644–688, http://www.jstor.org/stable/1339137.

Chengappa, R. (2000). *Weapons of peace: Secret story of India's quest to be a nuclear power*. New Delhi: Harpercollins.

Chidambaram, R. (2006). India and the International Atomic Energy Agency. *Atoms for Peace: An International Journal, 1*(2–3), 137–144.

Cooper, M. (2011). Nuclear liability: The market-based, post-Fukushima case for ending Price-Anderson. *Bulletin of the Atomic Scientists*. Retrieved October 17, 2012, from http://thebulletin.org/web-edition/features/nuclear-liability-the-market-based-post-fukushima-case-ending-price-anderson.

Currie, E. J. D. (2008). The problems and gaps in the nuclear liability conventions and an analysis of how an actual claim would be brought under the current existing treaty regime in the event of a nuclear accident. *Denver Journal of International Law and Policy*, *35*(1), 85–127.

Department of Energy & Climate Change, UK (2012). Implementation of changes to the Paris and Brussels Conventions on nuclear third party liability. A public consultation. Department of Energy & Climate Change, UK. Retrieved September 5, 2012, from http://www.decc.gov.uk/en/content/cms/consultations/paris_brussels/paris_brussels.aspx.

Edwards, R. R. (1957). The development of civil atomic energy programs in the free world. *Proceedings of the American Society of International Law (1921–1969)*, *51*, 149–155.

ElBaradei, M. (2007). Statement to the Fifty-First Regular Session of the IAEA General Conference, transcript, *IAEA*. Retrieved September 17, viewed March 22, 2010, from http://www.iaea.org/newscenter/statements/2007/ebsp2007n014.html.

Epstein, W. (1977). Why states go—and don't go—nuclear. *Annals of the American Academy of Political and Social Science*, *430*, 16–28.

European Commission. (2007). Communication from the commission to the council and the European Parliament–Nuclear Illustrative Programme {SEC(2007) 1261} {SEC(2007) 1262}, viewed 10 September 2012 from http://eur-lex.europa.eu/legal-content/EN/TXT/?uri=CELEX:52007DC0565.

Faure, M. G. & Vanden Borre, T. (2008). Compensating nuclear damage: A comparative economic analysis of the U.S. and International Liability Schemes. *William and Mary Environmental Law and Policy*, *33*(1–5), 219–286, viewed March 26, 2011, from http://scholarship.law.wm.edu/wmelpr/vol33/iss1/5

Fischer, D. (1997). *History of the International Atomic Energy Agency: The first forty years*. Vienna: IAEA.

Freeman, A. V. (1960). The development of international co-operation in the peaceful use of atomic energy. *The American Journal of International Law, 54*(2), 383–392.

French Embassy in New Delhi. (2013). Interview given by François Hollande, President of the Republic, to the Indian Newspaper, transcript, *The Times of India*. Retrieved February 15, 2013, viewed 20 February 2013, http://ambafrance-in.org/Interview-given-by-Francois.

Fuller, J. G. (1975). *We Almost Lost Detroit*. Retrieved March 25, 2011, from http://wsrl.org/pdfs/detroit.pdf.

Galiette, J. (1978). The Price-Anderson Act: A constitutional dilemma. *Boston College Environmental Affairs Law Review*, *6*(4), 565–596. Retrieved February 24, 2011, from http://lawdigitalcommons.bc.edu/cgi/viewcontent.cgi?article=1863&context=ealr.

Ginsburgs, G. (1960). The Soviet Union and international co-operation in the peaceful use of atomic energy bilateral agreements. *The American Journal of International Law*, *54*(3), 605–614. Retrieved March 10, 2011, from http://www.jstor.org/stable/pdfplus/2195311.pdf?acceptTC=true.

Green, H. P. (1973). Nuclear power: Risk, liability, and indemnity. *Michigan Law Review, 71*(3), 479–510.

Handrlica, J. (2010). Euratom powers in the field of nuclear liability revisited. *International Journal of Nuclear Law, 3*(1), 1–18.

Hardy, M. J. L. (1961). International protection against nuclear risks. *The International and Comparative Law Quarterly, 10*(4), 739–759.

Heiss, H. (1993). Legal protection against transboundary radiation pollution: A treaty proposal. *Fordham Environmental Law Report* (Vol 4, pp. 167–194).

Hinteregger, M. (1998). The new austrian act on third party liability for nuclear damage, *Nuclear Law Bulletin*, *62*(2), 27–50, viewed April 25, 2011, from http://www.oecd-nea.org/law/nlb/Nlb-62/hintereg.pdf.

Hunt, C. D. (1977). Canadian policy and the export of nuclear energy. *The University of Toronto Law Journal, 27*(1), 69–104.

IAEA. (1953). Atoms for peace speech, transcript *IAEA*. Retrieved March 27, 2012, from http://www.iaea.org/About/atomsforpeace_speech.html.

IAEA. (1960). Civil liability for nuclear hazards. *IAEA Bulletin, 2*(3), 8–11. Retrieved January 10, 2010, from http://www.iaea.org/Publications/Magazines/Bulletin/Bull023/02305000812.pdf.

IAEA. (1964). The Geneva conference: How it began. *IAEA Bulletin, 06*(3). Retrieved May 15, 2011, from http://www.iaea.org/Publications/Magazines/Bulletin/Bull063/06305100303.pdf.

IAEA, (1968). Safety, efficiency and economy of nuclear power in USSR. *IAEA Bulletin, 39*(15), 3–8. Retrieved June 15, 2012, from http://www.iaea.org/Publications/Magazines/Bulletin/Bul l102/10204700308.pdf.

IAEA. (1997). The 1997 Vienna Convention on Civil Nuclear Liability for Nuclear Damage and 1997 Convention on Supplementary Convention for Nuclear Damage-Explanatory Texts. Retrieved March 20, 2010, from http://wwwpub.iaea.org/MTCD/publications/PDF/Pub1279_web.pdf.

IAEA. (2006). Environmental consequences of the Chernobyl accident and their remediation: Twenty years of experience. *Report of the Chernobyl Forum Expert Group 'Environment', Radiological assessment reports series*. Retrieved May 23, 2013, from http://www-pub.iaea.org/mtcd/publications/pdf/pub1239_web.pdf.

IAEA. (2011). *International Fact Finding Expert Mission of the Nuclear Accident following the Great East Japan Earthquake and Tsunami*. Retrieved May 20, 2012, from http://www.iaea.org/newscenter/focus/fukushima/missionsummary010611.pdf.

IAEA. (2012). *IAEA Action Plan on Nuclear Safety—Nuclear Liability*. Retrieved September 10, 2012, from http://ola.iaea.org/OLA/documents/ActionPlan.pdf.

ICJ. (1949). Corfu channel (United Kingdom of Great Britain and Northern Ireland v. Albania), from http://www.icjcij.org/docket/files/1/1663.pdf

Kaplan, S. G. (1988). Compensating damage arising from global nuclear accidents: The Chernobyl situation. *Loyola of Los Angeles International and Comparative Law Review, 10*, 241–269. Retrieved June 20, 2012, from http://digitalcommons.lmu.edu/ilr/vol10/iss1/7.

Krige, J. (2006). Atoms for peace, scientific internationalism, and scientific intelligence. In J. Krige and K.-H. Barth (Eds.) *Global power knowledge: Science and technology in international affairs* (Vol. 21, pp. 161–181). Chicago: University of Chicago Press. Retrieved March 26, 2011, from http://www.jstor.org/stable/4129759.

Lamm, V. (1998). The protocol amending the 1963 Vienna Convention. *Nuclear Law Bulletin, 61*(1), 7–24.

Lamm, V. (2006). The Protocol amending the 1963 Vienna Convention. *International Nuclear Law in the post-Chernobyl Period*, viewed May 18, 2013, from https://www.oecd-nea.org/law/chernobyl/LAMM.pdf

Loewen, E. P. (2011). The history of probabilistic risk assessment of severe reactor accidents, transcript, *Sociedad Nuclear Mexicana LAS/ANS Symposium*. Retrieved 8 August, viewed September 18, 2012, from http://www.new.ans.org/about/officers/docs/8-aug-11_mexico_address_fs.pdf.

Madigana, D. J., Baumann, Z., & Fisher, N. S. (2012). Pacific bluefin tuna transport Fukushima-derived radionuclides from Japan to California. *Proceedings of the National Academy of Sciences, 109*(24), http://www.pnas.org/content/109/24/9483.full.

Mallison, W. T., Jr. (1957). The United States bilateral agreements for co-operation in the civil uses of atomic energy. *Proceedings of the American Society of International Law (1921–1969), 51*, 142–149.

Malone, L. A. (1987). *The Chernobyl accident: A case study in international law regulating state responsibility for transboundary nuclear pollution*. William & Mary Law School Faculty Publications. Pap. 590. Retrieved May 17, 2011, from http://scholarship.law.wm.edu/facpubs/590.

Marcus, G. H. (2008). The OECD nuclear energy agency at 50. *Nuclear News*, pp 27–33. Retrieved May 19, 2011, from www.new.ans.org/pubs/magazines/download/a_550.

Meehan, T. (2012). Lessons from the Price-Anderson nuclear industry indemnity act for future clean energy compensatory models. *Connecticut Insurance Law Journal, 18*(1).

Michal, R. (2001). Fifty years ago in December: Atomic reactor EBR-I produced first electricity, *Nuclear News*. Retrieved November 2001, viewed May 17, 2012, from http://www.ne.anl.gov/About/reactors/ebr1/2001-11-2.pdf.

Nadar, R., & Abbots, J. (1979). *The menace of atomic energy*. London: Stratford Books.

Nordykea, M. D. (1998). The Soviet program for peaceful uses of nuclear explosions. *Science and Global Security, 7*, 1–117. Retrieved August 10, 2012, from http://www.princeton.edu/sgs/publications/sgs/pdf/7_1nordyke.pdf.

OECD-NEA, *NEA 50 Anniversary (1958–2008)*, e-book. Paris: OECD. Retrieved August 20, 2011, from http://www.oecd-nea.org/general/reports/nea6365-50th-historical.pdf.

OECD-NEA. (2009). *Priority Rules on Compensation for Nuclear Damage in National Legislation*. Retrieved March 10, 2011, from http://www.oecd-nea.org/law/TABLE%20-%20Priority%20rules%2015%2012%2009.pdf.

OECD-NEA. (2012). *Fukushima Press Kit*. Retrieved December 14, 2012, from http://www.oecd-nea.org/press/press-kits/fukushima.html.

Osaka, E. (2012). *Liability for the Fukushima Nuclear Disaster*, viewed November 12, 2012, from http://digital.law.washington.edu/dspace-law/bitstream/handle/1773.1/1161/21PRPLJ433.pdf?sequence=1

Otway, H. J. (1974). Nuclear power plant safety. *IAEA Bulletin, 16*, 1–2. Retrieved September 22, 2011, from http://www.iaea.org/Publications/Magazines/Bulletin/Bull161_2/161_202007277.pdf.

Pelzer, N. (2010). Main features of the revised international regime governing liability-progress and standstill. In *International nuclear law: History, evolution and outlook*. Paris: OECD/NEA.

Petros'yants, A. M. (1984). A pioneer of nuclear power; Obninsk marks 30 years of nuclear power. *IAEA Bulletin, 26*(4). Retrieved May 26, 2012, from http://www.iaea.org/Publications/Magazines/Bulletin/Bull264/26404794246.pdf.

Prime Minister's Office, Japan. (2011). *Press Briefing for Members of the Foreign Press*. Retrieved 22 April, viewed May 20, 2012, from http://www.kantei.go.jp/foreign/incident/pdf/foreign-press-briefing-20110422-script-e.pdf.

Radetzki, M. (1999). Limitation of third party nuclear liability: Causes, implications and future possibilities. *Nuclear Law Bulletin, 63*(1), 7–24. Retrieved April 20, 2011, from http://www.oecd-nea.org/law/nlb/NLB-63/radetzki.pdf.

Raju, S., & Ramana, M. V. (2010). The other side of nuclear liability. *Economic & Political Weekly, xlv*(16), 48–54.

Ram Mohan, M. P. (2011). All dressed up and nowhere to go. *The Financial Express*. Retrieved December 26, viewed February 20, 2013, from http://www.financialexpress.com/news/all-dressed-up-and-nowhere-to-go/892022/0.

Reyners, P. (2013). A new world governance for nuclear safety after Fukushima? *International Journal of Nuclear Law, 4*(1), 63–77.

Sachs, N. (2008). Beyond the liability wall: Strengthening tort remedies in international environmental law, *UCLA Law Review*, *55*, 837–904.

Schwartz, A. J. (2006). International nuclear third party liability law: The response to Chernobyl. In *International nuclear law in the Post-Chernobyl period*. Paris: OECD.

Schwartz, A. J. (2010). Liability and compensation for third party damage resulting from a nuclear accident. In *OECD-NEA* (ed), *International nuclear law: History, evolution and outlook*, e-book. Retrieved January 15, 2011, from http://www.oecd-nea.org/law/isnl/10th/isnl-10th-anniversary.pdf.

Sethna, H. N. (1979). India's atomic energy programme—Past and future. *IAEA Bulletin*, 21(5), 2–11. Retrieved May 19, 2012, from http://www.iaea.org/Publications/Magazines/Bulletin/Bull215/21505090211.pdf.

Shils, E. (1948). The failure of the united nations atomic energy commission: an interpretation. *The University of Chicago Law Review, 15*(4), 855–876.

Soljan, V. (2000). The new definition of nuclear damage in the 1997 protocol to amend the 1963 Vienna Convention on civil liability for nuclear damage. *Reform of Civil Nuclear Liability, International Symposium Budapest*, Hungary 31 May–3 June 1999, OECD Nuclear Energy Agency. Retrieved April 09, 2012, from https://www.oecd-nea.org/law/legislation/nea2188-liability.pdf.

Sugimoto, A., Krull, S., & Nomura, S., et al. (2012). The voice of the most vulnerable: lessons from the nuclear crisis in Fukushima, Japan. *Bulletin of the World Health Organization, 90*(8), 629–630. Retrieved July 20, 2012, from http://www.who.int/bulletin/volumes/90/8/11-094474/en/index.html.

UN. (1972). General assembly resolution 2996 (XXVII). *International responsibility of States in regard to the environment*, viewed July 22, 2012, from http://daccess-ddsny.un.org/doc/RESOLUTION/GEN/NR0/270/26/IMG/NR027026.pdf?OpenElement.

U.S. Atomic Energy Commission (US AEC). (1957). *Theoretical Possibilities and Consequences of Major Accidents in Large Nuclear Power Plants 1957-WASH-740*. Retrieved January 28, 2011, from http://www.dissident-media.org/infonucleaire/wash740.pdf.

United States Atomic Energy Act of 1954, as Amended (P.L. 83–703).

United States Government Accountability Office. (2010). *Government wide Strategy Could Help Increase Commercial Benefits from U.S. Nuclear Cooperation Agreements with Other Countries*. Retrieved April 26, 2011, from http://www.gao.gov/new.items/d1136.pdf.

UNSEAR. (2008). Health effects due to radiation from the Chernobyl accident, *Report to the General Assembly with Scientific Annexes*, viewed August 30, 2012, from http://www.unscear.org/docs/reports/2008/11-80076_Report_2008_Annex_D.pdf.

UNSCEAR. (2012). *Interim Findings of Fukushima-Daiichi Assessment presented at the Annual Meeting of UNSCEAR*, United Nations. Retrieved October 11, 2012, from http://www.unis.unvienna.org/unis/pressrels/2012/unisous144.html.

US Congress. (1984). *Nuclear power in the age of uncertainty*, viewed on June 10, 2011, from http://www.fas.org/ota/reports/8421.pdf

US NRC. (1990). *Final Summary Report, Severe Accident Risks: An Assessment for Five U.S. Nuclear Power Plants- NUREG-1150*. Retrieved January 28, 2011, from http://www.nrc.gov/reading-rm/doc-collections/nuregs/staff/sr1150/v1/sr1150v1-intro-and-part-1.pdf.

US NRC. (2012). *Fact Sheet on Nuclear Insurance and Disaster Relief Funds*. Retrieved February 18, 2013, from http://www.nrc.gov/reading-rm/doc-collections/fact-sheets/funds-fs.html.

US Nuclear Regulatory Commission (US NRC). (1975). *Reactor Safety Study: An Assessment of Accident Risks in U.S. Commercial Nuclear Power Plants—WASH 1400*. Retrieved January 28, 2011, from http://pbadupws.nrc.gov/docs/ML0706/ML070610293.pdf.

Warren, G. C. (2000). Nuclear damage under the 1997 protocol: conventional thinking? *Reform of Nuclear Liability Law*. Retrieved March 25, 2010, from http://www.oecd-nea.org/law/legislation/nea2188-liability.pdf.

Willrich, M. (1965). The development of international law by the international atomic energy agency. *Proceedings of the American Society of International Law (1921–1969), 59*, 153–160.

Woodliffe, J. (1990). Chernobyl: Four years. *The International and Comparative Law Quarterly, 39*(2), 461–471.

Chapter 3
The Indian Civil Liability for Nuclear Damage Act, 2010: An Analysis

Abstract The making of the 'Civil Liability for Nuclear Damage Act 2010' was one of the finest legislative endeavours in the recent times. The exercise was significant because nuclear energy and the consequences of pursuing such an energy form were debated extensively in the Parliament for the first time. The result was a liability law that had an exceptional domestic political acceptability, but in many ways appeared to defy conventional international practice. The international nuclear community, led by supplier countries and vendors has argued that the law should be amended to be compatible with the established practice of international nuclear liability law. Examining through two specific examples—limitation of liability and right of recourse, the author argues, that though the Indian law gives the impression of defiance, the Parliament has only utilised the provisions of international nuclear law conventions—expanding boundaries of interpretation. Further, a section on transboundary applicability of the Indian law, and India's commitment under Convention on Supplementary Compensation to its neighbours is analysed to identify the operational difficulties.

Keywords India's civil nuclear liability law · Convention on supplementary compensation · Right of recourse · Limited liability · India's nuclear liability and transboundary application · International nuclear liability law

3.1 Introduction

The India–United States Civil Nuclear Energy Cooperation, initiated in 2005 (DAE 2005), was successfully concluded in 2008 with the Nuclear Suppliers Group's (NSG) waiver (IAEA 2008) to India, and the final approval by the US Congress. As India is not a member of the Nuclear Non-Proliferation Treaty (NPT), these engagements facilitated India's entry into the international civil nuclear trade after four decades of international isolation. With the conclusion of these international legal formalities, India is planning a major expansion of its

M.P. Ram Mohan, *Nuclear Energy and Liability in South Asia*,
DOI 10.1007/978-81-322-2343-6_3

civilian nuclear sector through both high-capacity imported reactors and an indigenous program. France, Russia and the US are already facilitating high-capacity reactor imports. The Government of India has specified that the expansion of the nuclear energy program will be through joint ventures, with majority share holding by the government owned and controlled, Nuclear Power Corporation of India (NPCIL) (Parliamentary Standing Committee 2010). Currently, under the *Atomic Energy Act 1962* (Atomic Energy Act or AEA), only the Central Government or government owned companies are allowed to undertake commercial production of nuclear energy (Section 3, AEA). Companies, both foreign and domestic, who become stake holders in the joint ventures, will have a pre-determined roles such as, technology and equipment providers, fuel suppliers, etc. In the foreseeable future, NPCIL will remain the sole 'operator' (Lok Sabha 2010).

It has been argued that the key benefit of *The Civil Liability for Nuclear Damage Act, 2012* (Act No 38 of 2010) (Civil Liability Act or The Act), has been to provide a predictable civil nuclear liability regime. The supplier countries demanded a set of liability rules based on the established principles of international nuclear liability law. A codified law is also an important tool in alleviating the concerns of the public.

Prior to the enactment of the Civil Liability Act, India had seen the evolution of the principles of liability through tort laws. The courts applied these principles to liabilities arising from dangerous and hazardous industrial activities. Especially in the post-Bhopal disaster scenario, the Supreme Court liberally interpreted and expanded the scope of the principles of liability. The concept of 'strict liability' evolved in the 1868 case of *Ryland vs. Fletcher* (L.R. 3 H.L. 330; [1861–73] All E.R) was further strengthened by the decision of the Supreme Court in *M C Mehta v. Union of India* (*Oleum Gas Leak Case:* AIR 1987 SC 1086); where the concept of 'absolute liability' was expounded and elaborated. It was in the *Oleum Gas Leak* case that the court stated 'an enterprise which is engaged in a hazardous or inherently dangerous activity that poses a potential threat to the health and safety of persons and owes an *absolute* and *non-delegable duty* to the community to ensure that no harm results to anyone'. The principle of absolute liability is operative without any exceptions, unlike the strict liability principle. The Supreme Court reiterated this principle in *Indian Council of Enviro-Legal Action v. Union of India* (AIR 1996 SC 1466), where the court observed that the industry alone has the resources to discover and guard against hazards, and dangers caused by its actions.

Such a liberal interpretation of the principles of liability by the Supreme Court was part of a wider trend of judicial activism demonstrated by the Indian Judiciary in the post-emergency era (the Government of India suspended Constitutional rights under Article 352 of the Constitution from 26 June 1975–21 March 1977). The most-striking feature of the activist phase was the expansion of Article 21 of the Constitution of India which guarantees the right to life and personal liberty. This expansion has occurred through the following two ways: (i) By subjecting a law restricting life and personal liberty to non-arbitrariness and reasonableness contained in Articles 14 and 19, respectively; and (ii) through the reading of unenumerated rights into Article 21 (Divan and Rosencrantz 2002). Through this interpretation, the right to environment became part and parcel of right to life, thereby giving wider powers to the Supreme Court and the High Courts to intervene in those areas, including liability issues arising from hazardous activities.

The flip side of the development of the principle of 'absolute liability' is that the same may be tantamount to 'unlimited liability'; obviously, a point of concern for the suppliers, because of the uncertainty the same creates in business. *The Environmental Protection Act, 1986* was enacted to provide adequate legal safeguards to protect and preserve the environment, but the law did not provide a compensation regime. In the aftermath of the Bhopal accident, *the Public Liability Insurance Act, 1991* was enacted to provided a legal remedy for immediate relief to the persons affected by an industrial accident. However, the PLIA specifically exempts 'war' and 'radioactivity' from its scope [Section 2(a)].

Therefore, with the opening up of nuclear technology commerce, India needed a liability regime that would be publicly acceptable; and also meet the industry's requirement of a regime based on the internationally accepted principles. The Civil Liability Act, attempts to address these issues. The Parliamentary and public debate concerning the Civil Liability Act focuses largely on three major areas; (1) The liability amount and limitation in liability, (2) right of recourse and (3) India's adherence to the Convention on Supplementary Compensation.

This chapter has not discussed in detail the concept of 'right of recourse' as the same does not directly concern the transboundary liability issues in South Asia. However, as this chapter exclusively focuses on the Indian legal position, the provisions relating to the right of recourse have also been discussed.

3.1.1 Making of the Civil Liability for Nuclear Damage Law

The Civil Liability Act, attempts to fulfill the needs of two different groups of stakeholders: The vendors and the public at large. While the vendors require certainty, the public wishes to be able to recover adequate compensation for all those who may be adversely impacted by an accident. As Varadarajan (2010a, b) states, 'the primary motive of the liability legislation in a democracy ought to be reassuring people that their interests would be fully looked after in the unlikely event of an accident'. For the industry, to obtain adequate insurance, they require clear identification of the extent of the liability. Thus the absence of a civil liability law was the last hurdle before the international suppliers undertook business in India. The Prime Minister's statement after the passing of the Bill in Lok Sabha captures the importance the government attached to the law, with him calling the same the 'Completion of a journey to end the apartheid against India in the field of atomic power' (Cherian 2010).

No doubt, the law has been framed so as to act as an incentive for attracting foreign nuclear technology investments. The Parliamentary debates on the Indo-US negotiations and the Civil Liability Act, represented the first substantial discussion on the nature of liability arising from accidents in the NPPs. The debate also left the polity deeply divided on the issue of nuclear energy (Rajgopalan and Purushothaman 2012).

The *Civil Liability for Nuclear Damage Bill, 2010* was introduced in the Lok Sabha on 7 May, 2010 and was referred to the Standing Committee on Science and Technology, Environment and Forests for detailed examination and submission of report (The Standing Committee). The Standing Committee after detailed examination presented the report on 18th August, 2010. The Committee was unanimous that the 'Bill being a domestic legislation should reflect Indian interests' (Standing Committee 2010). The report requested a number of changes to make the law stronger and for greater public acceptance. Though the Parliamentary Standing Committee recommendations were not binding; the recommendations were considered in good faith and were incorporated.

In the case of nuclear liability bill, most suggestions were incorporated. Important changes that had been recommended were raising the compensation amount to 1500 crores from 500 crores; the central government was empowered to take additional measures beyond the capped amount if the amount of compensation exceeds 300 million SDR; increasing the time limit for filing claims for injury from 10 to 20 years; provision of elaborate right of recourse provisions against the suppliers and upholding the victims' Constitutional right to approach the Supreme Court and the High Courts, other than the Claims Commissioner.

The Civil Liability for Nuclear Damage Bill was passed by the Parliament, and received the Presidential approval on 21 September 2010 (Act No 38 of 2010). After a year and 2 months, the Act was notified, coming into force on 11 November 2011. *The Civil Liability for Nuclear Damage Rules, 2011* (referred to as the 'Civil Liability Rules' or 'The Rules') have also been framed in respect of a few provisions, and were notified on the same day along with the Act.

The conformity of the Act and the Rules with the internationally accepted principles of nuclear liability law, however, is an issue that is yet to be settled. In 2010, India became a signatory to *the Convention on Supplementary Compensation for Nuclear Damage, 1997* (Compensation Convention or CSC), facing substantial criticism both domestically (Gopalakrishnan 2011; Ramachandran 2010 etc.) and internationally (Curtis 2010; Pelzer 2011). It was also reported that, the US had asked India to 'engage with the IAEA to ensure that the Indian nuclear liability law fully conforms to the Compensation Convention' (Dikshit 2011). In December 2012, a senior US government official said, 'India's nuclear liability law is not in line with the international nuclear liability principles reflected in the Convention on Supplementary Compensation for Nuclear Damage,' (PTI 2012). India has not attempted to pacify the concerns of the international suppliers through a modified interpretation of rules. Though the international suppliers led by the US argue for amendments to the law to harmonise it with the international principles, there is significant domestic pressure not to do so. The Indian Government remains committed to the law as it stands today. This is clearly reflected in the 2015 Frequently Asked Questions (FAQs) released by the Ministry of External Affairs (MEA), subsequent to discussions with the US government (MEA 2015). The FAQ states that "there is no proposal to amend the Act or the Rules". The US government now appears to appreciate the substance of law, and its interpretation by the Indian government. In February 2015, India and the US reached an agreement of understanding on the issues related to civil nuclear liability and this includes an understanding that Indian law is broadly compatible with CSC (MEA 2015). The Table 3.1 provides a comparison of the international principles and the enacted Indian law.

Table 3.1 The Indian law and the international principles: a comparison

International principles	Indian law
Absolute (strict) liability	No-fault liability. **Section 4(4)**
Legal channeling	Liability channeled to the operator [**Section 4(1)**]
	Provisions of the right of recourse (**Section 17**) and also the citizens' right under tort (**Section 46**)
Exclusive jurisdiction	The Nuclear Claims Commissioner. Constitutional right to approach the High Courts and the Supreme Court exists in parallel
Limited liability in amount	
Operator liability	**Section 6** The operators of the nuclear installations producing more than 10 MW of energy shall be liable up to Rs. 1500 crores
	Research and spent fuels re-processing plants will have different liability amounts
	Through a notification, the central government has the power to increase the operator's liability
Government liability	**Section 6 and 7** The total liability for a nuclear incident capped at 300 million Special Drawing Rights. Provisions for additional relief if the cap of 300 million SDR is insufficient
	The government may assume the liability of a nuclear installation by notification, if it feels that doing so is essential in the public interest
Other remedies for victims	**Section 46** Provides, that the provisions in the Act shall be, in addition to, and not in derogation of, any other law, for the time being in force, and nothing contained herein shall exempt the operators from any proceedings which might, apart from this Act, be instituted against such operators
Operator's right of recourse	**Section 17** The law provides that the operators after paying the compensation for nuclear damage shall have a right of recourse against the suppliers under the following conditions • such right is expressly provided for in a contract in writing; • the nuclear incident has resulted as a consequence of an act of the supplier or his employee, which includes supply of equipment or material with patent or latent defects or sub-standard services; • if the nuclear incident has resulted from an act of commission or omission of an individual done with the intent to cause nuclear damage
Limited liability in time	**Section 18** For damage to property, the time limit is 10 years
	For personal injury to any person, the time limit is 20 years
Financial security	**Section 8** All operators (except the central government) need to take insurance or provide financial security to cover their liability
	Currently, only the Central Government is allowed

Source Civil Nuclear Liability Act, 2010 Compiled by the Researcher

3.2 Analysis of the Civil Liability Act

A review of the law and the associated debates covers three main aspects. First, whether the law provides for a limited liability regime or can it be interpreted as an open-ended liability regime? Second, how does the implementation of the right of recourse work, and is this a deviation from an established practice? Third, the transboundary applicability. These issues are discussed in detail below.

3.2.1 Liability: Limited or Open-Ended?

The Paris Convention, 1960 and the Vienna Convention, 1963 provide for limited liability in amount and time. The Convention on Supplementary Compensation for Nuclear Damage, 1997, which is not yet in force also provides the same. This principle has been incorporated in multiple national legislations around the world, with varying amounts, and sharing of responsibility between the government and the operator in various forms. To cite a few examples, the US has limited operator liability and unlimited government liability; liability is limited for both the operator and the government in France; Germany has unlimited liability for the operator and limited liability for the government; the UK has limited liability regime for both the operator and the government. Table 3.2 lists the types of liability systems present internationally.

Under the Civil Liability Act of India, although the liability amount has been statutorily limited to 1500 crores, the law provides for the enhancement of the liability amount [Section 6(2)]. On a reading of the Act, and the position taken by the government during the parliamentary debates, it would be incorrect to state that the 'liability *per se*' is limited forever to Rs. 1500 crores. The operator's liability currently fixed at Rs. 1500 crores, and can be increased by a notification of the Central Government [Section 6(2) proviso]. Though, any increment of the amount is subject to a ceiling of 300 million Special Drawing Rights (approximately

Table 3.2 Types of liability internationally

Country	Operator liability	Government liability
United States	Limited	Unlimited
France	Limited	Limited
Japan	Unlimited	Unlimited
Russia	Unlimited	Unlimited
South Korea	Limited	Not mentioned
United Kingdom	Limited	Limited
Germany	Unlimited	Limited
India	Limited	Unlimited

Source Nuclear operator liability amounts and financial security limits. OECD/NEA (2014) and compiled by the Researcher

Rs. 2050 crore) [Section 6(1)]. Nonetheless, the proviso to Section 6(1) makes it clear that vide notifications, the ceiling of 300 SDR can also be raised.

This indicates that the government has kept its option open with regard to nature of limited liability. Moreover, it should be emphasised that neither the Atomic Energy Act, nor the Civil Liability Act allow a private enterprise to lead an NPP [Section 1(4) and Section (1) of AEA]. One of the key issues before the Parliamentary Standing Committee was the proposed role of private enterprises. The Standing Committee's response to the question was "The Bill applies only to nuclear installations owned and controlled by the Central Government either by itself or through any authority or corporation established by it or a government company, as defined in the Atomic Energy Act, 1962". The Committee suggested the inclusion of such a provision 'will make the position clear that there will be no private operator of nuclear installation' (Standing Committee 2010).

Currently, NPCIL is the only government company authorised for business of nuclear energy. NPCIL will hold a majority stake in all the joint ventures, both international and domestic. Therefore, since the Civil Liability Act applies solely to a Government Company [Section 6(1)]) and as the Central Government will take over the liability if it exceeds the amount specified in the Act [Section 7(1)]; the combined reading of the provisions leads to the conclusion that the operator liability can be termed as unlimited in India.

Internationally, the liability conventions in force today do not strictly limit liability. Either it is open to the member countries to legislate the extent of liability. Though the Paris Convention, through Article 7(a) provides for a maximum liability, Article 7(b)(i) States that 'any Contracting Party, taking into account the possibilities for the operator of obtaining the insurance or other financial security required pursuant to Article 10, may establish by legislation a greater or lesser amount'. In allowing the OCED countries to legislate as per their own requirement, in clearer terms, the Paris Convention in Article 15 further states, 'Any Contracting Party may take such measures as it deems necessary to provide for an increase in the amount of compensation specified in this Convention'. The 2004 Amended protocol and the Paris Convention retains the same 'intent' provisions with minor changes. The IAEA supported Vienna Convention on the other hand provides that 'the liability of the operator may be limited by the Installation State to not less than US $5 million' (the 1997 Protocol increased the same to not less than 300 million SDRs) for any one nuclear incident. The Vienna Convention is very specific that it is up to the Installation State (Operator State) to either limit or to provide for unlimited liability. The only requirement being, that at least 300 million SDR should be available as compensation, in case of an adverse event.

The Compensation Convention to which India is a signatory, and to which the Civil Liability Act adheres to, does not, in any way, limit the liability (Article 3). As in the case of the other conventions, the only requirement is that the domestic law of the country should specify a minimum amount. It is left to the States to provide for a maximum or 'greater amount' [Article 3(a)(1)]. As McRae (2007), suggests, "The CSC focuses on making an assured amount of compensation available. Accordingly, the CSC does not establish either a floor or a ceiling on the liability of the operator or require an Installation State to limit the liability of the operator". This means that an unlimited liability regime is possible within the CSC. Subsequent to India's ratification of the CSC, India may very well retain all its liability provisions that are said to be inconsistent. However, the provision of a greater amount is required to be provided from public funds. In India's case, since no private nuclear operators are conceived of or anticipated to act as operators, any financial consequences to an accident will have to be borne entirely by the government—through government-owned operators or by the government directly (Sections 6 and 7).

In fact, during the discussions of the Civil Liability Bill, there were strong political views that the amount of 1500 crores was itself too low. During the Parliamentary Debates, the Communist Party of India (Marxists), a left political party had suggested that the liability be increased to Rs. 10,000 crores (The Hindu 2010).

Nevertheless, the clear insertion [Section 6(1)] that this law applies only to the government owned and controlled companies, is itself a landmark achievement in the liability law discourse. More recently, the IAEA has supported the debate on the concern that the capping of liability in effect provides an additional subsidy to the operator (IAEA 1997; Currie 2008). It can be argued that the Civil Liability Act fully appreciates India's judicial precedent which expanded the concept of liability to 'absolute' and 'unlimited'. With the law providing for unlimited liability to be borne by the government, as in the Indian case, the possible magnitude of the liability would act as a deterrent against a lax legal regime. The law allows the operator to seek liability from errant suppliers through its right of recourse.

It is this exceptional right of the operator to seek damages from the supplier and subjecting them under Section 46 (tort claims) that has been a cause of concern with the suppliers (both domestic and international) (Curtis 2010; Srinivasan 2012). On account of such deviations, Pelzer (2011) completely misunderstood the nature and intent of Indian law when he says it deviates from the concept of legal channeling and therefore does not follow the international principles.

Section 46 provides, 'that the provisions in the Act shall be in addition to, and not in derogation of, any other law for the time being in force, and nothing contained herein shall exempt the operator from any proceeding which might, apart from this Act, be instituted against such operator'. Interestingly, this provision was hardly debated before the Standing Committee and in the Parliament. The Standing Committee acknowledged the constitutional right of Indian citizens to approach the Supreme Court and the High Courts under writ jurisdiction. This provision in the comparative nuclear liability discourse is unique. It is the Government of India's position that the section allows for the fixation of liability through both tort and criminal action, in addition to any other statutory claims. Abraham (2011) argues 'even when the right of recourse is not pursued by the operator, i.e. NPCIL under Section 46, any other affected individual or entity can seek tortious remedies against any supplier whose actions, equipment or materials are responsible for a nuclear incident'. However, the Indian Government in 2015, subsequent to a series of Contact Group meetings with US officials clarified that this Section does not expose Supplier or Vendors in any way to potential tort claims (MEA 2015).

In fact, a series of Public Interest Litigations have been filed across the country arguing that the capping of liability goes against the judicial interpretations. In the instant PIL before the Supreme Court, the court was requested to examine the constitutional validity of the Act that limits the liability of an operator to Rs. 1500 crores (Special Leave Petition (C) No. 27335 of 2012 along with SLP(C) 29121 of 2012 and WP(C) 407/2012). Petitioners, emphasising on the 'polluter pays principle' and 'absolute liability principle', argue that the capping of nuclear liability at Rs. 1500 crores is ultra vires of Article 21 of the Indian Constitution, which protects life and liberty. They further argue, that 'Nuclear equipments are expensive and even minor safety additions can easily exceed the maximum liability amount of Rs. 1500 crores as per the Act'. It is reported that the court observed, 'On the legalities of the Act, we are with you and we will examine the vires of the Act, but not on other issues. Whether a reactor is harmful, it is viable, it can be allowed or not, we do not have the expertise. They are all highly scientific issues. Can this court interfere?' (Venkatesan 2012). The Supreme Court is currently hearing the case. The final decision is awaited.

3.2.2 Operator's Right of Recourse: Indian Innovation or Nightmare in Interpretation?

One of the most contentious provisions of the Civil Liability Act pertains to the supplier's liability—the operator's right of recourse against the supplier. Internationally, one of the founding principles of the nuclear law, i.e. legal channeling of liability; provides that the operator is solely responsible for an accident. Nevertheless, the three international nuclear liability regimes allow the operator to seek remedies against its supplier under special circumstances (Table 3.3). The domestic laws of the nuclear power producing countries provide for:

(a) blanket immunity for the suppliers;
(b) full right to pursue unlimited claims against the supplier;
(c) a narrow right for the operator's to go for recourse.

The provisions relating to right of recourse under the Indian liability law are as follows:

> Section 17: The Operator of the nuclear installation after paying the compensation for nuclear damage in accordance with Section 6, shall have a right to recourse where:
> (a) Such right is expressly provided for in a contract in writing;
> (b) The nuclear incident has resulted as a consequence of an act of suppliers or his employees, which includes supply of equipment or material or patent or latent defects or sub-standard services;
> (c) The nuclear incident has resulted from the act of commission or omission of an individual done with the intent to cause nuclear damage

A reading of the above provision reveals that in the event of an accident, the operator is required to pay compensation first. After payment of compensation,

Table 3.3 Right of recourse—a comparison

Vienna convention, 1963	Only, (a) if it is fully expressed in writing, and (b) If the damage results from an act or omission done with intent to cause damage, against the person who caused the damage
Paris convention, 1961	Only, (a) against someone for an act of commission or omission with intent to cause damage and (b) If there is a clear contract giving such a right
Convention on Supplementary Compensation, 1997	National law may provide that the operator shall have a right of recourse only: 1. If this is expressly provided for by a contract in writing; or 2. If the nuclear incident results from an act or omission done with intent to cause damage, against the individual who has acted or omitted to act with such intent
Brazil, Canada and France	There is no provision giving a right against the supplier
Japan	Right of recourse exists against a third party causing damage
South Korea	Provides for recourse against the supplier in case of willful act or omission
Russia	No limit is imposed on the operator's right of recourse

Source Compiled by the Researcher, PRS Legislative Research, Legislative Brief: The Civil Liability for Nuclear Damage Bill, 2010 and Matveev (2006)

during the investigation, if the operator concludes that the accident occurred due to the fault of a supplier; the operator has a right of recourse against any such supplier. The section appears to provide an open-ended time limit for the operator to seek recourse. The question here is whether such a detailed right is an internationally accepted practice? Under the conventions, there is a restricted right and domestic jurisdictions have provided for either an absolute right; no right; or a very restricted right.

Clearly, the Indian law with the inclusion of Section 17(b), provides for a stronger domestic law, than that contemplated under in the Compensation Convention. Ramachandran (2011) reports that this expanded meaning seems to be inconsistent with the Compensation Convention. However, it can be argued that such extensive rights do exist in many domestic jurisdictions including, US and Russia. The US has objected to such a provision and now wants the IAEA to vet the Indian law. Varadarajan (2010a) points out that 'the US system of economic channeling of liability allows tort claims as well as an unrestricted right of recourse for the operator'. Further, to take a position, India can very well go beyond the international norms and strengthen its domestic legal regime, taking into account its domestic interests. The Indian law clearly provides the grounds based on which an operator can exercise its right of recourse against his supplier. Though it has been argued that the Indian law is not in consistence with some of the provisions of the CSC, mere strengthening of the law does not tantamount to illegality. In any case, such an extended right of recourse is a significant achievement in law making, considering the competing interests that were in play. Gruendel and Kini (2012) states that 'it is quite apparent that the discussions pertaining to the operator's right of recourse did not take the existing international liability regime as a starting point'. Hariharan (2011) views India's position as one in the right direction—reforming the outdated international nuclear liability regime. On the other hand, in practice, this extensive right may be difficult to implement, i.e. suing the supplier(s) who are likely be dispersed across the world.

The Civil Liability Rules 2011 have sought to clarify some of the interpretational issues. Chap. 5, Rule 24: Right of Recourse states:

> A contract referred to in clause (a) of Section 17 of the Act shall include a provision for the right of recourse for not less than the extent of the operator's liability under sub-Section (2) of Section 6 of the Act or the value of the contract itself, whichever is less.
>
> The provisions for right of recourse referred to above sub-rule (1) shall be for the duration of initial license issues under the Atomic Energy (Radiation Protection) Rules, 2004 or the product liability period, whichever is longer.

Rule 24 defines 'Product Liability Period' and 'Supplier'. 'Product Liability Period' means the period for which the supplier has undertaken liability for patent or latent defects or sub-standard services under a contract [Rule 24 Explanation 1(a)]. The term 'supplier' has been given a broad definition [Rule 24 Explanation 1(b)][1]. The operator according to the Rules is entitled to sue any or all the suppliers for damages under a 'right of recourse' claim. At the same time, the Rules also limits the claim against the supplier both in amount and time. The Rules restrict the right of recourse to either the duration of the initial license or the product liability period, whichever is longer. Five years is the period of the initial license, accordingly the liability is limited to 5 years [the Atomic Energy (Radiation Protection) Rules 2004: Rule 24(2) and Rule 9].

Many interpretations have been put forward regarding Rule 24 of the Civil Liability Rules. Sorabjee (Greenpeace 2010) in his opinion stated, "Rule 24(1) is clearly inconsistent with Section 6 of the said Act read with Section 17 inasmuch it scales down and reduces the liability prescribed by the said Act. Consequently the said proposed Rule is ultra vires the said Act and is invalid". Reacting to Sorabjee's opinion, Abraham (2011) argues, "Rule 24 would have absolutely no application in cases falling under Section 17(b) and (c), i.e. where the nuclear incident has resulted as a consequence of an act of a supplier or his employee which includes supply of equipment with patent or latent defects or sub-standard services

---

[1] Supplier shall include a person who—Manufactures and supplies, either directly or through an agent, a system, equipment or component or builds a structure on the basis of functional specification; or

Provides build to print or detailed design specifications to a vendor for manufacturing a system, equipment or component or building a structure and is responsible to the operator for design and quality assurances; or

Provides quality assurances or design services.

or if it is done by an individual with the intent to cause damage. Therefore, it would not be accurate to state that the supplier's liability has entirely been limited by virtue of Rule 24. In the event the circumstances under Section 17(b) or (c) are made out, the operator would clearly have a right of recourse against the supplier".

The Indian government was asked to clarify the position in the Lower House of the Parliament (the Lok Sabha). The question raised was, 'whether the new rules are in conformity with the Nuclear Liability Act passed by the Government or a dilution of the Act and violation of the spirit of the Act' (Lok Sabha 2011). In response the government stated 'The Civil Liability for Nuclear Damage Rules, 2011 are in conformity with the *Civil Liability for Nuclear Damage Act, 2010*' (Lok Sabha 2011). Leaving aside the interpretational issues, rules have revived the fundamental questions that were raised during the enactment of the parent Act itself. Sengupta and Ambast (2012) state that the Rule clearly limits the extended meaning of the right of recourse given under the parent Act. Gopalakrishnan (2011) observes that the Rule limiting the liability of the supplier is the result of Section 6(1) of the Act which caps the operator's liability. According to him, "This, unfortunately, is the consequence of Parliament limiting the operator's liability for nuclear reactor accidents at Rs. 1500 crore in Section 6(2) of the Liability Act, against the repeated pleas at that time to substantially increase that cap. In my view, the only option now left is to argue for higher compensation, including consequential damages, under Section 46, relying on Tort Laws". The Chairman of the Subordinate Legislation Committee of the Lok Sabha that reviewed the Rules for the conformity with the parent Act, also reportedly remarked on the issue of the quantum and period of compensation; stating that the Rules have gone beyond the intent of the Act, which in effect limits the supplier's liability (Jigeesh 2012).

On the whole, the enactment of the Civil Liability Act represents the Parliament's concerns regarding fixation of responsibility. However, the execution of the law, in particular the explanation made under the Rules seems to have created interpretational issues. There is a marked divergence between the official positions as stated in the Parliament, that the Rules are in conformity with the Act; and opinion of legal experts and other political parties who find the Rules to be circumventing the parent Act.

3.3 The Indian Law, the Compensation Convention and Transboundary Applicability

The issue of transboundary applicability of the Indian law *vis-a-vis* India's commitment under the Compensation Convention was hardly discussed within and outside the Parliament. The Parliamentary Standing Committee, in its review of the Civil Liability Act, briefly touched upon the issue and noted "the geographical scope of damage caused by a nuclear accident may not be confined to national boundaries and it may have transboundary effects" (Standing Committee 2010). Further, during the discussion the report articulated:

> Some of the experts opined that the CSC stipulates that the compensation arising out of nuclear damages, made available by other CSC members, can be spent only to compensate the damages suffered in the territory of a 'Contracting Party', namely, a country which itself has acceded to the CSC. Accordingly, if an Indian nuclear accident causes transboundary adverse impacts in any of the neighbouring countries, the liability for compensation has to be made from the Indian taxpayers' funds, if no supplementary financial assistance from the CSC is forthcoming.

Other than these statements, there has not been any major discussion on the issue in the Parliament or in any other forum.

The Civil Liability Act extends to the whole territory of India and its maritime zones, including vessels registered in India.[2] India has also made commitments to an international liability convention, becoming a signatory to the Convention on Supplementary Compensation. Importantly, before the Standing Committee (Standing Committee 2010) and during the Parliamentary debates (Lok Sabha 2012), the Government of India has taken a position that India's law are in tune with the Annex of the Compensation Convention. The effect of this position, and India's accession to the Compensation Convention has implications on transboundary applicability. As mentioned above, India has not yet ratified the Convention. It comes into effect in April 2015. The coming into force of the treaty is likely to bring to fore issues pertaining to transboundary liability and compensation, and India's financial obligations relating to the same.

When India ratifies the CSC, what does it mean for India? Will it allow India to access international funds to pay victims within the country; and will this also make India responsible for the payment of compensation to neighbouring countries?

The Compensation Convention is a free standing instrument, separate from the other existing liability regimes—the Paris and Vienna Conventions. The CSC mandates that the compensation for nuclear damage be distributed equitably; without discrimination on the basis of nationality, domicile or residence [Article 3 (2)(a)]. The Convention also states that in case of damages suffered by a non-contracting party neighbour, the Compensating State is free to exclude such damages for compensation. The one exception to this position is provided by Article XII(2), which

---

[2] Section 1(2) It extends to the whole of India.

(3) It also applies to nuclear damage suffered—(a) in or over the maritime areas beyond the territorial waters of India

(b) in or over the exclusive economic zone of India as referred to in Section 7 of the Territorial Waters, Continental Shelf, Exclusive Economic Zone and Other Maritime Zones Act, 1976; 80 of 1976.

(c) on board or by a ship registered in India under Section 22 of the Merchant Shipping Act, 1958 or under any other law for the time being in force; 44 of 1958.

(d) on board or by an aircraft registered in India under clause (d) of sub-Section (2) of Section 5 of the Aircraft Act, 1934 or under any other law for the time being in force; 22 of 1934.

(e) on or by an artificial island, installation or structure under the jurisdiction of India.

states that such exclusion shall not affect compensation amount to a State that is a non-nuclear installation State solely on the basis of lack of reciprocity (IAEA Explanatory Texts Section 3.5.3).

In the Indian and South Asian context this means, that for the Convention to operate in South Asia, every country should either have laws consistent with the CSC or it should be a member of the Vienna Convention. The Paris Convention is not referred here since it does not apply to this region. In the Indian case, ratification and entry into force of the Convention would make India liable to other Contracting Parties and would also be required to contribute to an international fund (Articles 6 and 7, CSC). Though the Convention requires adherences from all the generating and non-generating countries; exceptionally in the case of non-nuclear States, the Convention states that the Parties should not withhold compensation *solely* on basis of lack of reciprocity [Article XII(2)]. Thus, if an accident in India impacting Nepal or Bhutan or Sri Lanka–non-generating and also non-members, the interpretation of provisions of the Convention obligates India to pay compensation. On non-contracting states, the Convention does not specify any bar to the payment of compensation, but has left it to the Contracting States to decide as per their discretion [Article III(2)(a)].

On the question of availability of funds, the Compensation Convention assures compensation for nuclear damage in the member countries by providing for a two tier system of compensation. The Convention requires the Contracting Parties to have a minimum amount which is pegged as 300 million SDR or provide for a greater amount. This is the same amount as has been specified in the Indian law [Section 6(1)]. The Convention further states, that in case the operator's funds are insufficient to cover the first tier amount (minimum of 300 million SDR), then CSC requires the Installation State to make public funds available to cover the difference (Article 1 CSC). As a second tier, in case the claims exceed the 300 million SDR, the Compensation Convention requires the member countries to contribute to an international fund (Articles 6 and 7, CSC). This international fund is not fixed, 'but rather is dependent on the number of nuclear power plants in member countries and will increase as the number of such plants increase' (McRae 2007). According to Articles 3(1)(b) and 5, the use of international funds is only within the geographic scope of the Convention and does not cover non-member countries. The international fund is distributed in the following ways, (a) 50 % of the funds shall be available to compensate the claims for nuclear damage suffered in or outside the Installation State; and (b) 50 % of the funds shall be available to compensate the claims for nuclear damage suffered outside the territory of the Installation State to the extent that such claims are uncompensated under (a) above (Article XI). Moreover, the Convention encourages the parties to have a three-tier compensation structure which is solely managed by the agreed parties (Article XII).

With respect to the availability of funds and its distribution, the Convention is clear that it should be based on non-discrimination of nationality, domicile or residence [Article 3(2)(a)]. This means that with the ratification of the CSC, when the convention comes into effect, the amount referred in Indian law, i.e. 300 million SDR, in the event of transboundary consequence, has to be distributed between the affected citizens and the neighbours. The Indian victims will not have any preferential treatment. Only in the case of a non-contracting party, a refusal could be made. In the case of South Asia, except Pakistan, most of the countries are either non-generating countries or soon to become members of the Vienna Convention. Being a contracting State to the CSC, India will have access to this international fund. This fund, the extent of which is open for interpretation, can be used for providing compensation both within and outside the country. An additional worry in respect to adherence is that this Convention specifically provides that all the parties should ratify, accept or approve the Compensation Convention. Article XVIII provides that the Vienna Convention or the Paris Convention States should also deposit the instrument of ratification, approval or acceptance.

The hope of a harmonised regime within South Asia led by the CSC, will materialise only when all the countries in the region, regardless of nuclear power programs or membership to other conventions, become party to it. Such a complex adherence may not seem to be practical.

The debates within India barely touched India's transboundary commitment through the CSC or otherwise. The right of recourse and its interpretational issues became the dominant legal discussion. Being a member of the CSC, in the event of a transboundary accident, Indian victims may not be fully compensated from first tier compensation due to its non-discriminatory commitments. The Government may have to take over the additional funding of compensation; to satisfy the Indian victims and for the reparation of environment—this would involve payment by the tax payers.

3.4 Conclusion

The Civil Liability Act has raised more questions than answers. It is clear that the limited liability regime that has been advocated as an integral part, may not be a case in actuality. Apart from the perceived limitation imposed by the statute, the statute has given extraordinary right for the public to seek 'unlimited' compensation. The clarification from the government, vide the 2015 MEA FAQs, demonstrates the government's commitment to the law and also eagerness to move forward with nuclear energy collaboration. This has assuaged some of the concerns of suppliers. However, it is very likely that the application of the law, other than the Constitutionality which is already being heard in court, will be challenged before courts. Thus, the certainty sought still remains elusive. In respect of the right of recourse, the operator, i.e. NPCIL has been vested with enormous authority and responsibility; even after dilution suggested through the Rules. However, the implementation of the joint venture agreements between NPCIL and the supplier countries, and its interpretation will require clarification, considering that there exist a lot of confusion on the exact meaning and nature of the Rules framed. In respect of the argument of inconsistency of the Indian law with the CSC, a position can be taken that the Indian approach is fully in line with the evolution of international nuclear liability law, i.e. importance of domestic law in the evolution of the CSC. On the question of transboundary applicability, once India ratifies the CSC, provision of compensation payment is based on the principle of non-discrimination. Since there is no priority principle either in the domestic law or the CSC, India would be asked to treat all the victims, wherever they reside, irrespective of country and offer compensation.

References

Abraham, M. (2011). *Right of Recourse: Interpretation under Civil Liability for Damage Liability Rules, 2011*. Retrieved November 19, 2011, from http://nuclearlaw.wordpress.com/2011/11/19/right-of-recourse-interpretation-under-civil-liability-for-damage-liability-rules-2011/.

Cherian, J. (2010). Safety last. *The Frontline, 27*(19), 98–100 (Frontline).

Currie, E. J. D. (2008). The problems and gaps in the nuclear liability conventions and an analysis of how an actual claim would be brought under the current existing treaty regime in the event of a nuclear accident. *Denver Journal of International Law and Policy, 35*(1), 85–127.

Curtis, L. (2010). India's flawed nuclear legislation leaves U.S.–India partnership short. *The Heritage Foundation*. Retrieved July 24, 2012, from http://www.heritage.org/research/reports/2010/08/indias-flawed-nuclear-legislation-leaves-us-india-partnership-short.

DAE. (2005). *Joint Statement on India-USA Civil Nuclear Co-operation*, viewed May 10, 2010, from http://www.dae.nic.in/?q=node/61.

Dikshit, S. (2011). U.S. wants IAEA to vet Indian liability law, *The Hindu*. Retrieved 20 July, viewed October 24, 2012, from http://www.thehindu.com/todays-paper/us-wants-iaea-to-vet-indian-liability-law/article2260867.ece.

Divan, S., & Rosencrantz, A. (2002). *Environmental law and policy in India* (3rd ed.). New Delhi: Oxford University Press.

Gopalakrishnan, A. (2011). *Why the Nuclear Liability Rules Need to be Modified, DNA*. Retrieved December 5, viewed November 23, 2013, from http://www.dnaindia.com/analysis/comment-why-the-nuclear-liability-rules-need-to-be-modified-1621411.

Greenpeace. (2010). *Soli Sorabjee Opinion on Civil Nuclear Liability*. Retrieved June 25, 2010, from http://www.greenpeace.org/india/en/publications/open-letter-from-greenpeace/soli-sorabjee-opinion-on-civil/.

Gruendel, R. J., & Kini, R. E. (2012). Through the looking glass: Placing India's new civil liability regime for nuclear damage in context. *Nuclear Law Bulletin, 1*(89), 45–66.

Hariharan, A. (2011). India's nuclear civil liability bill and supplier's liability: One step towards modernizing the outdated international nuclear liability regime. *William & Mary Environmental Law and Policy Review, 36*(1), 223–255.

IAEA. (1997). *The 1997 Vienna Convention on Civil Nuclear Liability for Nuclear Damage and 1997 Convention on Supplementary Convention for Nuclear Damage—Explanatory Texts*, viewed March 20, 2010, from http://www-pub.iaea.org/MTCD/publications/PDF/Pub1279_web.pdf.

IAEA. (2008). *Communication dated 10 September 2008 received from the Permanent Mission of Germany to the Agency regarding a "Statement on Civil Nuclear Cooperation with India"*, INFCIRC/734 (Corrected), IAEA. Retrieved April 27, 2012, from http://www.iaea.org/Publications/Documents/Infcircs/2008/infcirc734c.pdf.

Jigeesh, A. M. (2012). Govt pulled up for diluting Civil Nuclear Liability Rules, *Business Line*. Retrieved August 28, 2013, from http://www.thehindubusinessline.com/industry-and-economy/govt-pulled-up-for-diluting-civil-nuclear-liability-rules/article3832551.ece.

Lok Sabha. (2010). Private sector participation in nuclear power sector, *Unstarred Question no. 413*. Retrieved January 18, 2012, from http://dae.nic.in/writereaddata/lsus240210.pdf.

Lok Sabha. (2011). New rules for civil nuclear liability, *Unstarred question no. 2228*. Retrieved June 2013, from http://dae.nic.in/writereaddata/2288-lsus071211.pdf.

Lok Sabha. (2012). Nuclear liability act, *Unstarred Question No. 2932*. Retrieved June 2013, from http://dae.nic.in/writereaddata/lsus2932.pdf.

Matveev, A. (2006). The Russian approach to nuclear liability. *International Journal of Nuclear Law, 1*(3), 270–286.

McRae, B. (2007). The convention on supplementary compensation for nuclear damage: Catalyst for a global nuclear liability regime. *Nuclear Law Bulletin, 79*(1), 17–36.

MEA. (2015). *Frequently Asked Questions and Answers on Civil Liability for Nuclear Damage Act 2010 and related issues*. Retrieved March 25, 2015, from http://www.mea.gov.in/press-releases.htm?dtl/24766/Frequently_Asked_Questions_and_Answers_on_Civil_Liability_for_Nuclear_Damage_Act_2010_and_related_issues.

OECD/NEA. (2014). *Nuclear operator: Liability amounts and financial security limits*, viewed March 30, 2015, from https://www.oecd-nea.org/law/2014-table-liability-coverage-limits.pdf.

Parliament of India, Department-related Parliamentary Standing Committee on Science & Technology, Environment & Forests 2010, *212th report—"The Civil Liability for Nuclear Damage Bill, 2010"*. Retrieved June 16, 2011, from http://www.prsindia.org/uploads/media/Nuclear/SCR%20Nuclear%20Liability%20bill.pdf.

Pelzer, N. (2011). The Indian civil liability for nuclear damage act, 2010—legislation with flaws? *ATW—International Journal for Nuclear Power, 56*(1).

PTI. (2012). India's liability law not in line with international norms: US, *The Hindu* 1 December. Retrieved December 30, 2013, from www.thehindu.com/news/international/indias-liability-law-not-in-line-with-international-norms-us/article4153366.ece.

Rajagopalan, R. P., & Purushothaman, U. (2012). Role of the Indian political system in shaping India's nuclear policy. *International Journal of Nuclear Law, 3*(4), 246–258.

Ramachandran, R. (2010). A supplier's deal? *Frontline, 27*(24) (Nov 20–Dec 03).

Ramachandran, R. (2011). Limiting liability. *Frontline, 28*(25) (Dec 03–16).

Sengupta, A., & Ambast, S. (2012). A dangerous recourse? A critical relook at section 17 of the civil liability for nuclear damage act, 2010. *Journal of International Nuclear Law, 3*(4), 292–307.

The Hindu. (2010). *Raise cap in nuclear liability Bill to Rs.10,000 crore: Left*, viewed August 24, 2013, from http://www.thehindu.com/todays-paper/tp-national/raise-cap-in-nuclear-liability-bill-to-rs10000-crore-left/article569571.ece.

Srinivasan, M. R. (2012). A liability for our nuclear plans, *The Hindu*. Retrieved October 15, viewed October 15, 2012, from http://www.thehindu.com/opinion/lead/a-liability-for-our-nuclear-plans/article3997102.ece.

Varadarajan, S. (2010a). Turn the nuclear bill from liability to asset, *The Hindu*. Retrieved June 16, viewed January 19, 2011, from http://www.thehindu.com/opinion/lead/article458314.ece?homepage=true.

Varadarajan, S. (2010b). This is no way to write a law, *The Hindu*. Retrieved August 24, viewed February 17, 2011, from http://www.thehindu.com/opinion/lead/article592116.ece?homepage=true.

Venkatesan, J. (2012). Supreme Court to examine constitutional validity of nuclear civil liability law, *The Hindu*. Retrieved March 17, viewed May 25, 2012, from http://www.thehindu.com/todays-paper/tp-national/supreme-court-to-examine-constitutional-validity-of-nuclear-civil-liability-law/article3004778.ece.

Chapter 4
A Nuclear Liability Framework for South Asia

Abstract South Asia is one of the densely populated regions of the world. A disaster in the nature of nuclear accident in one country will have a significant impact on the life and livelihood of large population across the region. Currently, major economies in South Asia are expanding their nuclear energy programmes, and this poses a transboundary risk. The risk is aggravated by the fact that countries in South Asia are not a part of any common international nuclear liability framework, nor do they have reciprocal domestic law. This subjects the region to an uncertain liability and compensation regime. This chapter explores the legal response mechanisms available in respect to state liability and compensation. The paper argues that the South Asian Association for Regional Cooperation (SAARC), a regional community of South Asian countries, is the appropriate institutional mechanism available to form a regional nuclear risk community.

Keywords Nuclear energy plans in South Asia · Nuclear energy liability and South Asia · SAARC regional cooperation on nuclear liability law · Transboundary impacts and nuclear risk · Nuclear power · Nuclear accidents · Compensation

4.1 Introduction

All over the world, particularly in the post-Chernobyl nuclear accident scenario, counties have pursued their nuclear energy programmes on the twin assurance of (1) utmost safety of the nuclear programme and, (2) the existence of adequate legal and institutional mechanisms to deal with state responsibility and subsequent liability claims, in the event of an accident involving radiation exposure (national or transboundary). The safety features of the nuclear power programmes refer to technological developments, the response and mitigation mechanisms, subsequent to an accident and the risk associated with the entire programme, can be secured only through strong legal and institutional measures. This chapter deals with the latter issue—the existence of legal and institutional structures with respect to a transboundary situation in the South Asian region.

M.P. Ram Mohan, *Nuclear Energy and Liability in South Asia*,
DOI 10.1007/978-81-322-2343-6_4

The South Asian countries—Afghanistan, Bangladesh, Bhutan, India, the Maldives, Nepal, Pakistan, and Sri Lanka, specifically the major economies, are now actively pursuing nuclear power programmes. India and Pakistan are set to expand their nuclear capacity substantially, Bangladesh and Sri Lanka will be the new entrants. Interestingly, all these developments are taking place without substantial public debate, nationally or regionally, and in the absence of any transboundary legal remedies and institutional structures.

South Asia with a population of 1.64 billion people represents 24 % of the world's population (UN Population Statistics 2011; Bloom and Rosenberg 2011). The region is also one of the biodiversity rich regions of the world (UNEP South Asia Environment Outlook 2009). As a densely populated, geographically linked region with rich biodiversity, South Asia will face serious adverse impacts on life and livelihood, in case of a major nuclear disaster. South Asia has a history of discord, this adds to the seriousness of the risk dimension, and adequate mitigation response to a nuclear accident. To deal with such eventualities and to make the countries accountable; cooperation and multilateral agreements; commitments between the countries are essential—both for an effective response mechanism and to secure remedies through adequate liability and compensation regime. Such an accident scenario, however remote, necessitates precaution and preparedness. Formation of a 'nuclear energy risk community' within South Asia is one way of addressing this shared concern.

The objective of such a nuclear risk community is to ensure that the South Asian countries, under the aegis of the South Asian Association for Regional Cooperation (SAARC) (South Asia and SAARC are used interchangeably), are capable of responding effectively as a community to a nuclear accident. This chapter argues that SAARC established in 1985, as a regional forum with memberships of all the eight South Asian countries is the appropriate forum to initiate a regional cooperation agreement on nuclear energy risk. There are initiatives already undertaken to assess the possibility of nuclear risk reduction strategies from nuclear weapons in South Asia (Centre for Strategic and International Studies 2004 and India Pakistan Agreement on Reducing Risk from Accidents Relating to Nuclear Weapons 2007 and Ottawa Track-two Dialogue 2011). These initiatives are solely concerned with threat reduction from nuclear weapons between India and Pakistan. None of the other South Asian countries had any major role or stake in these dialogues, as other than India and Pakistan, no other country in South Asia possesses a nuclear weapon programme. Unlike the discussion on nuclear weapons between India and Pakistan; nuclear energy, which is now being pursued by other countries within the SAARC, offers a more realistic multilateral approach to deal with the transboundary consequences of a nuclear accident.

This study deals with only the civilian nuclear energy programmes and its transboundary consequences, and is in no way concerned with the nuclear weapon programmes and its associated deliberations. The chapter examines the nuclear energy programmes, plans and policies of different SAARC countries; analyses the legal obligations and adherence to various nuclear conventions, and discusses the regional concerns in respect to liability and compensation. The SAARC's programmes, charters and mandates are critically reviewed to argue that SAARC is an appropriate forum to establish a 'SAARC nuclear energy risk framework cooperation'.

4.2 Nuclear Energy Programmes and Plans Within South Asia

The International Atomic Energy Agency (IAEA 2009a, b) projects that Asia, on account of massive energy requirements to fuel the economic progress and to meet the demands of a large population, may well be one of the engines of world's nuclear energy growth. Even though this projection was revised subsequent to the Fukushima accident in Japan, a major portion of the global expansion of nuclear power is still projected to be in Asia (IAEA 2011a). In South Asia, many countries have plans either to considerably expand, or to initiate domestic nuclear energy programmes. Country-specific nuclear energy programmes and plans are explained below.

4.2.1 India

India is the major nuclear energy player in South Asia. The nuclear programme in India is conceived as a unique sequential three-stage programme, essentially envisaged to use thorium, an abundantly available resource in the country, in the final fuel cycle (Chidambaram 2006). Internationally, the nuclear programmes do not rely on a close fuel cycle, and the spent fuel at the end of the first stage is treated as waste. Currently, India has an installed civilian nuclear capacity of approximately 4780 MW. India is not a party to the 1970 Nuclear Non-Proliferation Treaty (NPT) and also not a member of the Nuclear Suppliers Group (NSG). India being a non-signatory of the NPT, does not access nuclear technology from, and could not cooperate with the other NPT member countries in developing its nuclear energy programmes. In 2005, Government of India entered into a Civil Nuclear Cooperation agreement with United States (DAE 2005) aimed at ramping up its nuclear capacity to achieve energy targets. The successful negotiation of the agreement will allow India to import Uranium and high-capacity reactors to expand the nuclear power programme considerably.

Subsequent to the signing of the Civil Nuclear Cooperation Agreement in 2005, India successfully negotiated with the Nuclear Suppliers Group (NSG) countries for India's specific exception in respect of nuclear commerce (IAEA 2008); and also with the IAEA for India specific safeguards (IAEA 2009a–c, 2010). As of today, India has bilateral agreements with the major nuclear supplier countries—France, Russia, South Korea, UK, US, etc., and is negotiating with more supplier countries for nuclear resources, equipments and technology. Now that India can import technology and fuel, it has projected an ambitious nuclear energy plan. From approximately 4800 MW, the Department of Atomic Energy (DAE) plans to add 30,000 MW by 2020 and 60,000 MW by 2032 (Banerjee 2011). This capacity addition is planned through both the indigenous three-stage programme and importing high-capacity reactors through joint ventures.

With respect to international projects, the DAE has already sanctioned sites to joint-venture consortiums between NPCIL and foreign contractors. The selected sites are Haripur in West Bengal, MithiVirdi in Gujarat, Jaitapur in Maharashtra, Kovvada in Andhra Pradesh and Kudankulam in Tamil Nadu. Based on the information provided by the DAE in the Parliament, Table 4.1, details the proposed sites and the total capacity from both indigenous and imported reactors and Table 4.2 describes the sites for foreign collaborative projects, its total capacity and present status (Rajya Sabha 2012a).

4.2.2 Pakistan

Pakistan, another country in South Asia with nuclear energy capability, has also plans to expand its nuclear energy capacity significantly. Currently, Pakistan generates 462 MW annually from its nuclear programmes. In 2011, Pakistan Atomic

Table 4.1 Operational, under construction and proposed sites

States	Site	Capacity (MW)		
		Operational	Under construction	Proposed
Maharashtra	Tarapur	2 × 160 + 2 × 540	–	–
	Jaitapur[a]	–	–	2 × 1650
Rajasthan	Rawatbhata	100 + 200 + (4 × 220)	2 × 700	–
	Mahi, Banswara[a]	–	–	2 × 700
Tami Nadu	Kalpakkam	2 × 220	1 × 500	2 × 500
	Kudankulam	1 × 1000	1 × 1000	2 × 1000
Uttar Pradesh	Narora	2 × 220	–	–
Gujarat	Kakrapar	2 × 220	2 × 700	–
	Chhaya Mithi Virdi[a]	–	–	2 × 1100
Karnataka	Kaiga	4 × 220	–	2 × 700
Haryana	Gorakhpur[a]	–	–	2 × 700
Madhya Pradesh	Chutka[a]	–	–	2 × 700
	Bhimpur[a]	–	–	Pre-project activities
Andhra Pradesh	Kovvada[a]	–	–	2 × 1500
West Bengal	Haripur[a]	–	–	Pre-project activities

[a]New sites

Source Rajya Sabha (2012a), table updated by the author

Table 4.2 Foreign collaboration projects

Site	State	Capacity (MW)	Present status
Kudankulam	Tamil Nadu	4 × 1000[a]	Land available, environmental clearance from MoEF obtained, financial sanction of KK 3 and 4 obtained
Jaitapur	Maharashtra	6 × 1650	Land title transferred to NPCIL, environmental and CRZ clearances from the MoEF obtained, discussions with M/s. Areva on project proposal for JNPP 1 and 2 in progress
Kovvada	Andhra Pradesh	6 × 1000[b]	Land acquisition proceedings in progress, the ToRs for the EIA studies approved by the MoEF, the EIA studies in progress
Chhaya Mithi Virdi	Gujarat	6 × 1000[b]	Land acquisition proceedings in progress, the ToRs for the EIA studies approved by the MoEF, the EIA studies in progress
Haripur[c]	West Bengal	6 × 1000	Pre-project activities initiated

[a]The first pair KK 2 (1 × 1000 MW) attained criticality in 2013 and second is expected to be commissioned in 2015–2016
[b]Nominal capacity
[c]The Haripur project is currently shelved. The Government of West Bengal is refusing to accord permission to this project
Source Rajya Sabha (2012b), table updated by the author

Energy Commission (PAEC) announced it aims to achieve an installed capacity of 8800 MW by 2030 (Parvez 2011), with the expectation that China or possibly the US will provide assistance in its expansion plans (Bukhari and Rehman 2011).

Pakistan started its nuclear power programme by installing a 137 MW Canadian Deuterium Reactor (CANDU) at Karachi, Sindh in 1971, which became operational in 1972. According to Siddiqui and Qureshi (2005), the Indian nuclear test in 1974 had an adverse international effect on Pakistan as well. Subsequent to the Indian test, Canada in 1976 withdrew its post-contract technical support for the Karachi Nuclear Power Plant (KANUPP). In 2002 KANUPP completed its original design life of 30 years and the life extension programme of the plant is reported to have been started (Siddiqui and Qureshi 2005).

In 2002, in Chashma, Punjab; the country's second nuclear power plant of 300 MW started commercial operations. Construction of another unit of 300 MW at Chashma started in December 2005, which was inaugurated for commercial operation in May 2011. Two additional units, Chashma-3 and Chashma-4, are under construction at the same site and are expected to be commissioned by 2016–2017 (IAEA PRIS 2014).

Like India, Pakistan is also a non-member of the NPT. NPT bars non-member to participate in nuclear technology transfer even for peaceful purposes. Shah (2010) and Bukhari and Rehman (2011) report that Pakistan is lobbying for a nuclear cooperation agreement, either with the US or China or both. If such an agreement materialises, the international nuclear trade restrictions will be removed and a substantial increase in its plans for developing the nuclear energy sector can be expected.

Further, recent news reports, quoting an unnamed Pakistan Atomic Energy Commission official (Press Trust of India 2012) suggest that the government has identified new sites at the following locations—Taunsa-Punjnad canal, about 32 km away from Muzzafargarh, Punjab; Qadirabad Headworks, Punjab; Dera Ghazi Khan Canal near Taunsa Barrage, Punjab; Taunsa-Punjnad Canal near Multan, Punjab; Nara Canal near Sukkur, Sindh; Pat Feeder Canal near Guddu, Sindh and Kabul River near Nowshera, Punjab.

It is further reported that the Prime Minister of Pakistan "pledged to increase nuclear power generation capacity to 40,000 MW in the long term as part of his energy plan" (The Daily Times 2013). In June 2013 the Planning Commission of Pakistan in a statement said that two China National Nuclear Corporation 1100 MWe class reactors would be used for Karachi 2 and 3 (KANUPP 2 and 3). Following this, Executive Committee of the National Economic Council, Pakistan in July 2013 approved this plan (World Nuclear Association 2014). Further, the Chairman of the Pakistan Atomic Energy Commission Dr. Ansar Parvez stated that the work on five more plants of 1100 MW each would commence in the next 10 years. "The process for site selection of plants is continuing and with the passage of time indigenisation is increasing", he said (Syed 2014).

4.2.3 Bangladesh

Bangladesh has been pursuing its nuclear energy dream since 1961. Several feasibility reports have affirmed the technical and economic feasibility of a nuclear power plant at Rooppur. It is reported that even though in 1980, formal approval was granted for a 125 MW nuclear power plant (NPP), the plan has not yet materialised. Faced with a severe energy shortage in 2000 that culminated in riots, Bangladesh adopted the National Nuclear Power Action Plan (NNPAP). Under the NNAP, the country commits itself to a nuclear power programme in the future. Carrying forward the NNPAP, the 2004 National Energy Policy (NEP) (Ministry of Power, Energy and Mineral Resources 2004) identifies nuclear energy as one of the energy sources that is to be actively pursued.

Begum (2008), from the Bangladesh Atomic Energy Commission, states that the revised National Energy Policy of 2007 has identified establishment of two medium-size nuclear power plants by 2015 and 2017. The NEP 2007 proposes that 15–20 % of the national energy needs to be met from nuclear energy within 2 years (Begum 2008). Going further, Rahman (2012) states that the 2010 Power System Master Plan has projected a more ambitious nuclear energy target. As per the Power System Master Plan, the projection of nuclear power generation are two 1000 MW NPPs at Rooppur by the year 2020 and two 1000 MW plants by the year 2025 and 2030, respectively (see Table 4.3). Thus the aim is to travel from nil production to 4000 MW by 2030.

Table 4.3 Projection of the nuclear power generation—master plan 2010

Site	Capacity (MW)	Expected year
Rooppur 1[a]	1000	2020
Rooppur 2[a]	1000	2020
NPP3[a]	1000	2025
NPP4[a]	1000	2030

[a]All proposed: Rooppur 1 is currently being finalised
Source Information compiled from articles published by Begum (2008) and Rahman (2012)

Towards realising this long-term plan, Bangladesh and Russia have signed a framework agreement on the peaceful use of nuclear energy on 21 May 2010 (Rahman 2012). In November 2011, Bangladesh signed an intergovernmental agreement with Russia for the establishment of two nuclear power plants with a capacity of 1000 MW each at Rooppur (Haque 2009; IAEA 2011b). Taking it further, during October 2013, preliminary site works were started, and construction of the first unit is expected from 2015. Interestingly, Bangladesh is planning to have more nuclear power plants. It is reported that BAEC in 2014 invited the Japan Atomic Energy Agency (JAEA) to explore the possibility of building a second 2000 MW nuclear power plant in the south of the country (Rasel 2014). The same newspaper quotes State Minister for Science and Technology, "We will send our team to carry out survey in many isolated coastal islands in the country's southern part" as part of identifying a suitable site.

4.2.4 Sri Lanka

Sri Lanka is a new entrant in the nuclear power scenario. Sri Lanka, in recent times has taken a policy decision to incorporate nuclear power into its energy mix. Chapika Ranawaka, the Minster of Power and Energy of Sri Lanka, addressing the 54th Regular Session of the IAEA General Council in 2010, said, "Government has given clearance to set up a 1000 MW NPP, and has submitted a request to the International Atomic Energy Agency for a pre-feasibility study" (Ranawaka 2012). Further, Ranawaka argued that the intent of the pre-feasibility study is to have a nuclear energy option available beyond 2020.

4.2.5 Nepal

In 2007, the Government of Nepal took a major policy step towards developing nuclear science capabilities with the adoption of National Nuclear Policy (NNP) (Ministry of Environment, Science and Technology 2007). The NNP recognises the importance of the application of nuclear energy/technology in the development of Nepal.

As of today, Nepal does not have plans to establish a nuclear power programme. The policy is largely in respect to capacity building of human resources, and to use nuclear science and technology for medical, agriculture and other allied applications. Nevertheless, the policy is not silent in respect of a possible nuclear energy scenario. The background note of the policy clearly states in the following words:

> In the context of making maximum use of nuclear science in basic research in sectors of food and agriculture, health, energy, industry, geo-science, geo-chemistry, geo-physics and pollution measurement, it is useful for Nepal, too, to achieve better results on the above by utilizing nuclear technology for peaceful purposes.

On the future of nuclear power programmes, the policy provides;

> Importance of nuclear technology has increased since it is environmentally friendly and would be a better alternative to minimize the short supply of mineral fuels and to reduce severe environmental impacts caused by large scale hydroelectric power projects.

Afghanistan, Bhutan and the Maldives—the other three countries in the SAARC currently do not have any plans to develop a nuclear energy programme. Overall, the major players in South Asia are intensifying their efforts, and others have clearly indicated their plan to develop nuclear power.

4.3 International Legal Obligations on Civil Nuclear Energy and South Asia

South Asia is a special case in the international nuclear legal order, having the distinction of hosting two, of the four, non-member States to the NPT, i.e. India and Pakistan. Both the countries have energy as well as weapon programmes. A critical discussion on the reasons for India and Pakistan being outside the NPT is not the subject of this book. The study is limited to the regional implications within the SAARC with respect to nuclear energy development. Afghanistan, Bangladesh, Bhutan, the Maldives, Nepal and Sri Lanka are all parties to the NPT, and are eligible to receive nuclear technology for peaceful applications within the IAEA safeguards framework. It is under these provisions, Bangladesh and Sri Lanka are engaging with the IAEA and the supplier countries towards realising their nuclear energy options.

In order to facilitate the India–US nuclear deal, the US government amended its tough domestic law, *the Nuclear Non-proliferation Act, 1978* (NNPA 1978) along with relevant export licensing requirements. These export licensing requirements were framed in the aftermath of India's first nuclear test. Further, in order to accommodate India in the global nuclear commerce, US convinced the NSG members to amend the NSG guidelines, regarding nuclear export controls.[1] This is the route Pakistan may have to take, for India-like civil nuclear energy cooperation. On a positive note, all the countries in South Asia are contracting parties to the IAEA Emergency Conventions, which is discussed below. Table 4.4 shows the SAARC nations adherence to international nuclear legal instruments.

4.3.1 The IAEA Emergency Conventions and South Asia

Subsequent to the Chernobyl accident, as briefly discussed in Chap. 2, under the aegis of the IAEA, the Convention on Early Notification of a Nuclear Accident 1986 (Early Notification Convention) and the Convention on Assistance in Case of a Nuclear Accident or Radiological Emergency 1986 (Assistance Convention) (together referred as the Emergency Conventions) were adopted to provide for immediate notification and assistance in case of a nuclear accident. The Emergency Conventions establishes a framework to facilitate the exchange of information, and the prompt provisions of assistance in the event of a nuclear accident or radiological emergency. In case of an accident, specific obligations are placed on the parties and the IAEA, with the aim of minimising consequences to health, property and the environment.

The Convention on Early Notification of a Nuclear Accident, in Article 2 and 3, requires that in the event of a nuclear accident which has transboundary radiological consequences, the State Party has to immediately notify the neighbouring countries and the IAEA. Article 5(1) provides a detailed list of information to be provided about the accident. The Assistance Convention on the other hand requires that the Member States should cooperate amongst themselves and with the IAEA in order to facilitate prompt assistance, in the event of a nuclear accident or radiological emergency. This is to minimise its consequences and to protect life, property and the environment from the effects of radioactive releases (Article 1). The IAEA is empowered to promote, facilitate and support the cooperation between the States Parties (Article 2). Article 5 details the role of the IAEA in the event of a nuclear accident or radiological emergency. They are in summary:

- make available to a State Party or a Member State requesting assistance, appropriate resources for the purpose of conducting an initial assessment of the accident;

[1] NSG Guidelines **INFCIRC/254, Part I: Guidelines for nuclear transfers** The first set of NSG Guidelines governs the export of items that are especially designed or prepared for nuclear use. These include (i) nuclear material, (ii) nuclear reactors and equipment therefore, (iii) non-nuclear material for reactors, (iv) plant and equipment for reprocessing, enrichment and conversion of nuclear material and for fuel fabrication and heavy water production; and (v) technology associated with each of the above items. **INFCIRC/254, Part II: Guidelines for transfers of nuclear-related dual-use equipment, materials, software and related technology** The second set of NSG Guidelines governs the export of nuclear-related dual-use items and technologies, that is items that can make a major contribution to an unsafeguarded nuclear fuel cycle or nuclear explosive activity, but which have non-nuclear uses as well, for example in industry.

- transmit requests for assistance and relevant information to the States Parties that may possess the necessary resources; offer its good offices to the States Parties or Member States; and
- liaise with relevant international organisations to obtain and exchange relevant information; and, on request, coordinate the assistance at the international level that becomes available.

Importantly, however, the Emergency Conventions do not provide any legal obligations for liability and compensation claims. The liability and compensation claims from a nuclear damage are dealt under separate international nuclear liability conventions and also through domestic laws. All the SAARC countries have ratified the Emergency Conventions, albeit with reservations by India and Pakistan. Reservations by India and Pakistan provide an insight to the contrasts in the South Asian region—the deep mistrust between the neighbours.

On the substantive commitments under the Convention, both India and Pakistan have reservations with respect to the dispute settlement provisions (Article 11, para 2) (Declarations/reservations made upon expressing consent to be bound and objections thereto to the Convention on Early Notification of a Nuclear Accident 1986). This provision provides for the submission of disputes for arbitration or to the International Court of Justice at the request of any party to such dispute. Both India and Pakistan have submitted that for the submission of any international dispute for arbitration or to the International Court of Justice, the consent of all the parties concerned, in each individual case, is necessary.

In an accident scenario, if a country refuses to undertake its obligation, the Convention requires that, at the first instance, disputes should be settled through negotiations. If negotiations fail, 'any' party may refer the dispute for arbitration. However, the reservations by India and Pakistan provide that disputes cannot be referred for arbitration solely on the request of one party.

With respect to the Assistance Convention as well, India and Pakistan (as in the Early Notification Convention) have reservations on the substantive provisions of the Convention. India states, it does not consider itself obliged to accord privileges and immunities to assisting personnel (Article 8, paras 2 and 3). Further, India declares, that it does not consider itself bound to indemnify the assisting party from any legal proceedings; neither will it assume full responsibility of the proceedings and provisions for compensation to assisting parties in respect of death (Article 10, para 2). Regarding dispute settlement provisions, India has the same reservation as in the case of the Notification Convention (Article 13, para 2).

Pakistan too does not consider itself bound by paras 2 and 3 of Article 8, and also with respect to cases of gross negligence by the individuals who caused the death, injury, loss or damage (Article 10, para 2). Further, on the settlement of disputes, the reservations placed concern for submission of disputes for arbitration or to the International Court of Justice at the request of any party to such a dispute. Pakistan declares that for the submission of any international dispute for arbitration or to the International Court of Justice, the consent of all the parties concerned, in each individual case, is necessary (Article 13, para 2).

India and Pakistan's reservations to Article 10.2 with respect to the provisions of privileges and immunities, legal assistance, ownership of legal responsibility and compensation for death or injury of the 'assisting' party have far reaching consequences. The effect of the same is that in case of an accident, India and Pakistan would only be bound to notify each other and would not be fully obligated to accept the terms of the assistance convention. Therefore, assistance will be dependent on the then political and diplomatic situations. This is insensitive to the needs of the people of the region and undermines the treaty's purpose.

4.3.2 Emergency Conventions Dispute Settlement and International Court of Justice

India and Pakistan have similar reservations to the dispute settlement procedures of both the Emergency Conventions. These reservations in effect make the implementation difficult and dispute settlement cumbersome between the neighbours. Since there have not been many successful examples of bilaterally negotiated settlement of disputes between these two countries, the applicability of the convention itself is in doubt.

Even if one party approaches the International Court of Justice (ICJ), jurisdictional competence of the ICJ will have to be established.[2] The Court's jurisdiction rests on the consent of both the parties. Even compulsory jurisdiction of the ICJ

---

[2] Basis of the ICJ's Jurisdiction:
(a) *Special agreement*: Article 36, para 1, of the Statute provides that the jurisdiction of the Court comprises all cases which the parties refer to it. Such cases normally come before the Court by notification to the Registry of an agreement known as a *special agreement* and concluded by the parties specially for this purpose 2. The subject of the dispute and the parties must be indicated (Statute, Article 40, para 1; Rules, Article 39).
(b) *Cases provided for in treaties and conventions*: Article 36, para 1, of the Statute provides also that the jurisdiction of the Court comprises all matters specially provided for in treaties and conventions in force. In such cases a matter is normally brought before the Court by means of a written *application instituting proceedings* 3; this is a unilateral document which must indicate the subject of the dispute and the parties (Statute, Article 40, para 1) and, as far as possible, specify the provision on which the applicant founds the jurisdiction of the Court (Rules, Article 38).
(c) *Compulsory jurisdiction in legal disputes*: The Statute provides that a State may recognize as compulsory, in relation to any other State accepting the same obligation, the jurisdiction of the Court in legal disputes.
(d) *Forum prorogatum*: If a State has not recognised the jurisdiction of the Court at the time when an application instituting proceedings is filed against it, that State has the possibility of accepting such jurisdiction subsequently to enable the Court to entertain the case: the Court thus has jurisdiction as of the date of acceptance in virtue of the rule of *forum prorogatum*.
(e) *The Court itself decides any questions as to its jurisdiction*: Article 36, para 6, of the Statute provides that in the event of a dispute as to whether the Court has jurisdiction, the matter shall be settled by the decision of the Court. Article 79 of the Rules lays down the conditions which govern the filing of preliminary objections.
(f) *Interpretation of a judgment* (Rules, Article 98) 5 and (g) *Revision of a judgment* (Statute, Article 61, para 1).

does not imply mandatory settlement of disputes between state parties by the ICJ. States have the option of accepting the Court's jurisdiction, and can do so under terms and conditions they determine themselves (Alexandrov 2006).

Moreover, there have been instances where the Court has ruled that it has the jurisdiction; however, the same was contested by a party, who refused to implement the Court's final decision. In the Case *Concerning Military and Paramilitary Activities in and against Nicaragua*, US refused to participate in the proceedings and blocked the implementation of the decision of the ICJ. Scott and Carr (1987) declared this stance of the US as unfortunate. They wondered, "how the US, presumably a law-abiding and law-respecting nation, could possibly bring itself to snub the ICJ and, by extension, the ideal of international law".

There have already been instances wherein India and Pakistan have questioned the jurisdiction of the ICJ. The *Aerial incident of 10th August 1999 between Pakistan v. India* is a case in point (ICJ. Reports 2000, p. 12). On 21 September 1999, Pakistan filed an application in the ICJ instituting proceedings against India in respect of a dispute relating to the destruction of a Pakistani aircraft. Pakistan referred the dispute to the compulsory jurisdiction of the ICJ, while India based on its declarations/reservations to the ICJ's jurisdiction argued that the Court has no jurisdiction in this case (ICJ 2000). India invoked its reservations on the ICJ jurisdiction in disputes concerning the interpretation or application of a multilateral treaty, unless all the parties to the treaty give full jurisdiction to the ICJ.

The ICJ by fourteen votes to two found that it had no jurisdiction to entertain the application filed by Pakistan. The Court reaffirmed its view that "the judicial settlement of international disputes is simply an alternative to the direct and friendly settlement of such disputes between the States, which the Court is to facilitate" (Bekker 2000).

To sum up, in terms of practical implementation, the reservations put forward by India and Pakistan in respect of the Emergency Conventions and also the past experience with the ICJ points to the enormous difficulty in seeking a third party supervised settlement of disputes. Friendly settlement of disputes is a crucial element of peace and trust building between the countries. South Asian countries need to realise that a nuclear accident is a shared regional concern, and to deal with such an eventuality requires regional efforts.

4.4 South Asia and Nuclear Liability Laws

Internationally, the *1963 Vienna Convention on Civil Liability for Nuclear Damage* (the Vienna Convention) and the *1997 Convention on Supplementary Compensation* (CSC) specify the responsibilities of a State in case of liability and compensation claims. None of the countries in South Asia are party to the Vienna Convention, and barring India, no country has shown their interest in the CSC.

Table 4.4 The SAARC country wise legal regime and international commitments

SAARC nations	NPP	NPT	Notification convention	Assistance convention	Vienna convention	CSC	Domestic law
Afghanistan	×	✓	✓	✓	×	×	×
Bangladesh	✓ Planning stage	✓	✓	✓	× Domestic law in line with Vienna Convention	×	✓
Bhutan	×	✓	✓	✓	×	×	×
India	✓	×	✓	✓	×	✓ Not ratified	✓
Maldives	×	✓	✓	✓	×	×	×
Nepal	×	✓	✓	✓	×	×	×
Pakistan	✓	×	✓	✓	×	×	×
Sri Lanka	✓ Planning stage	✓	✓	✓	×	×	×

India signed the CSC on 27 October 2010, but has not ratified as of today. The signing signifies India's last commitment to the US stemming from the landmark 2005 nuclear agreement with the US.

At present there is no legal or treaty obligation upon the South Asian countries in respect of transboundary liability and compensation—a situation similar to the pre-Chernobyl liability framework. Other than treaty, another way of achieving consensus is to provide legal reciprocity. For example the United States *Price-Anderson Nuclear Industries Indemnity Act*, 1957 (42 USC §2210) and the Canadian *Nuclear Liability Act, 1985 as amended* (Nuclear Liability Act R.S.C., 1985, c. N-28) provide reciprocity of legal remedy towards liability and compensation in accident cases involving transboundary radiation.[3] Under the law of the respective country, regulations are in place: (1) *Declaring the United States to be a Reciprocating Country for Purposes of the Act* (SI/78-179), and, (2) *Canada-United States Nuclear Liability Rules* (C.R.C., c. 1240).

However, such a legal setup does not exist within the SAARC. At the domestic level, India has enacted *The Civil Liability for Nuclear Damage Act, 2010* (Act 38 of 2010), and Bangladesh in May 2012 enacted *The Bangladesh Atomic Energy Regulatory Act, 2012*. In respect of a transboundary scenario or transnational application, India's liability law is silent.

[3] Regulation made under Nuclear Liability Act (R.S.C., 1985, c. N-28): (1) Declaring the United States to be a Reciprocating Country for Purposes of the Act (SI/78-179), and, (2) Canada-United States Nuclear Liability Rules (C.R.C., c. 1240).

Bangladesh with the help of technical advisory provided by the IAEA, has finalised its regulatory and liability law. An official from the Bangladesh Atomic Energy Commission during a personal discussion observes that the Bangladesh law is in line with the Vienna Convention (18 February 2012). Table 4.4 provides a listing of SAARC country-specific nuclear laws, and adherence to international treaties and conventions.

To sum up, South Asia, though in the grip of nuclear power expansion, has not seen any concerted regional efforts in addressing transboundary nuclear liability concerns.

4.4.1 Transboundary Nuclear Liability Concerns Within the SAARC

In case of a nuclear accident with a transboundary impact, the situation as it stands today, allows a country to easily reject its responsibility and refuse payment of any compensation on account of its non-obligation under international nuclear law. This legal vacuum is compounded by the fact that, many countries within the SAARC are choosing the nuclear energy path, thereby increasing the risk quotient.

Interestingly, there were already strong voices on the transnational concerns from the NPPs within the SAARC. The Kudankulam Nuclear Power Project (KKNPP)[4] in Tamil Nadu, India is an example. At the time of setting up of the KKNPP, due to its proximity to the project; Sri Lanka voiced concerns that an accident could have an adverse impact on its people and environment (Rajya Sabha 2012c). Nepal, on its part, through its National Nuclear Policy 2007 also raised issues regarding the possible risk associated with large scale development of nuclear power projects in its neighbourhood. The National Nuclear Policy 2007 of Nepal, referring to transboundary concern, articulates the same in clear terms:

> Nepal being surrounded by several nuclear power plants in its neighbourhood and in the context of their number being continuously increased, Nepal cannot remain untouched from any kinds of nuclear accidents that may occur in its neighbourhood and therefore, it is urgent to make necessary measurements on the present status of radiation level and there should be regular monitoring and evaluation, and proper safety measures in this regard should be taken. Moreover, there is an urgent need to well inform the general public regarding the benefits and consequences of nuclear science.

The National Disaster Management Authority of India while drafting the nuclear emergency guidelines, "National Disaster Management Guidelines—Management of Nuclear and Radiological Emergencies, 2009" was conscious of the transboundary risk. The guidelines offer a passing reference (reproduced below), that

---

[4] Russia assisted Kudankulam Atomic Power Project will have in the initial phase 2*1000 MWe VVER-1000 reactors which is expected to be generate power by 2011/2012. Construction of another two reactors of the same specification is also begun on the same site. It is reported that the plant has the capacity to accommodate a total of 8 units.

in case of any nuclear accident from India or from the neighbouring countries, the concerned country will have to act in accordance with the country's international commitments—references to the IAEA Notification and Assistance Conventions.

> 8.5.4 Trans-Boundary Emergency: In accordance with international commitments, information will be provided to the concerned authorities/agencies in the event of transboundary effects of a major incident. In the event of the competent authority in the country receiving information about an incident elsewhere, which might have the potential of radiological impact within India, the National Nuclear/Radiological Emergency Plan would be activated.

It would be interesting to see whether India will have any reservations to Bangladesh's Rooppur NPP which is only 50 km aerially away from West Bengal. The responsibility of the offending State consequent to an accident, and reparation of damage has not yet been the subject of any discussion.

Pakistan in 2008 enacted, the "Regulations on Management of a Nuclear or Radiological Emergency". The regulation defines "significant transboundary release means a release of radioactive material to the environment that may result in doses or levels of contamination beyond national borders from the release which exceed international intervention levels or action levels for protective actions, including food restrictions and restrictions on commerce". The regulation further defines "transnational emergency as a nuclear or radiological emergency of actual, potential or perceived radiological significance for more than one State". This is qualified by stating that a significant transnational radioactive release does not mean a transboundary release. Though both the regulations of India and Pakistan do not expand State obligation, and liability and compensation issues consequent to a transboundary impact, these are to be seen as significant developments.

With respect to South Asia, the fundamental questions that had been raised in the 1960s and 1970s, when the nuclear energy programmes were still in a nascent stage, still remain unanswered. International nuclear law has developed and it holds the answers to these questions; the South Asian region is yet to gain control of the situation. The questions as summarised by Schwartz (2006) are:

> In the event of a nuclear incident occurring at or in connection with a nuclear installation in any one of these countries, victims who suffer injury, whether inside or outside of that country's borders, are likely to find themselves in a precarious position when it comes to claiming damages resulting from that incident—the same sort of precarious position in which Chernobyl victims found themselves. Do they have a right to sue? Who should they sue? Which courts will have jurisdiction to hear their claims? What is the burden of proof? How much money will be available to satisfy their claims? How quickly must they institute those claims? Where will the money come from to satisfy their judgments, assuming they are successful? Answering these questions involves a myriad of complicated legal, political and practical problems, many of which are, of course, resolved by the international nuclear liability conventions themselves.

Moreover, the concerns within South Asia stem not only from the lack of a legal framework, but also from the conflicting liability frameworks under development. The South Asian countries (party and non-party to the NPT), in the future are likely to accede to different international liability frameworks, while opting for different standards in domestic liability frameworks. By way of example, India is a party to the CSC; Bangladesh and Sri Lanka may become a party to the Vienna Convention due to the nature of engagement with the IAEA; Pakistan may also accede to the CSC in case it can secure an '*India-like*' nuclear deal. Interestingly, India's adherence to the CSC is itself contested by many countries due to non-conformity to the CSC Annex. The steps involved in achieving a wider harmonisation through the CSC—being the umbrella legislation, may be too onerous for the prospective State parties. The requirements as it stands are that (i) the CSC has to come into force and (ii) the countries that are party to the Vienna Convention must again ratify their adherence to the CSC as well.

A cause for concern is that the SAARC region would be without a harmonised regional framework. Adherence by the countries to one or the other international convention ends up defeating the purpose of adherence itself—harmonisation. Since none of the international liability regimes are applicable in South Asia as of today and there are no domestic liability laws for reciprocal arrangements, the region possibly will find itself in a quagmire of legal and political issues in case of an accident. It appears that the SAARC as a regional forum seems to be the only suitable institution with the capacity to obtain a consensus for the formation of a 'nuclear risk community'.

4.5 The SAARC's Establishment and Functions

The South Asian Association for Regional Cooperation (SAARC), a complex regional arrangement had its beginning as early as 1947. Dash (1996) observes that the idea of a regional cooperation in South Asia was discussed in at least three conferences: the Asian Relations Conference in New Delhi in April 1947, the Baguio Conference in the Philippines in May 1950, and the Colombo Powers Conference in April 1954. Its current avatar is credited to late President of Bangladesh, Ziaur Rahman who on 2 May 1980, first proposed the establishment of a regional cooperation framework in South Asia. It took several years of negotiation and many compromises including removal of contentious issues like security matters, internal political issues amongst others to establish SAARC. The association's founding and Charter (available at http://saarc-sdmc.nic.in/pdf/charter.pdf) were signed in Dhaka on 8 December 1985. The SAARC Charter, according to Obino (2009) "clearly defies the option of using the association as a direct political lever in order to resolve (either or both) bilateral and contentious issues as banned; according to Article X, these issues are to be kept clearly apart from the association's activities; also, the principle of unanimity was established as the only official decisions procedure, setting aside any institutional processing of dissent and disagreement".

The SAARC Charter is emphatic that the basis of cooperation among members is sovereign equality, territorial integrity, political independence and non-interference in internal affairs, and mutual benefit (SAARC Charter Article II, Principles No. (1). Further, the Charter also provides that such cooperation will not be an obstacle to other bilateral or multilateral cooperation or be inconsistent with them (SAARC Charter Article II, Principles No. (2).

4.5.1 SAARC Charter and Objectives

The SAARC Charter comprehensively lists the aspiration of the SAARC community, which are as follows (edited):

1. *Desirous of promoting peace, stability, amity and progress in the region through strict adherence to the principles of the UNITED NATIONS CHARTER and NON-ALIGNMENT, particularly respect for the principles of sovereign equality, territorial integrity, national independence, non-use of force and non-interference in the internal affairs of other States and peaceful settlement of all disputes*;
2. *Conscious that in an increasingly interdependent world, the objectives of peace, freedom, social justice and economic prosperity are best achieved in the SOUTH ASIAN region by fostering mutual understanding, good neighbourly relations and meaningful cooperation among the Member States which are bound by ties of history and culture*;
3. *Convinced that regional cooperation among the countries of SOUTH ASIA is mutually beneficial, desirable and necessary for promoting the welfare and improving the quality of life of the peoples of the region*;
4. …

And, the objectives of the Association states,

(a) *to promote the welfare of the peoples of SOUTH ASIA and to improve their quality of life;*
(b) …
(c) *to contribute to mutual trust, understanding and appreciation of one another's problems;*
(d) *to promote active collaboration and mutual assistance in the economic, social, cultural, technical and scientific fields;*
(e) ….

With respect to policy making and implementation, SAARC is organised into four structural levels. The first, the Council of Ministers consisting of the Foreign Ministers of the Member States as the apex policy making body of SAARC. The primary function of the Council of Ministers is the formulation of the policies; review of the progress of cooperation; decision on new areas of cooperation;

establishment of additional mechanism; and decision on other matters of general interest to the Association (Charter, Article IV). It is at the level of Council of Ministers that major decisions are taken. The Secretariat functions as a nodal office in following up the proposed mandates and is also instrumental in organising meetings. Within the ambit of SAARC's structure, in the past it has been successful in creating regional Agreements and Conventions.

The second is the Standing Committee consisting of Foreign Secretaries (Charter, Article V). The function of Standing Committee includes, overall monitoring and coordination of programme of cooperation; and importantly for the purpose of this research, identification of new areas of cooperation based on appropriate studies. The third and fourth are implementation committees, i.e. Technical Committee and Action Committee, constituted to implement country specific and cross-country joint projects, respectively (Charter, Articles VI and VII).

4.5.2 SAARC's Mandate and Formation of Nuclear Energy Risk Community

SAARC was conceived as an association to achieve South Asian regional integration. According to the SAARC Charter, the integration is to be achieved through mutual trust, understanding and appreciation of each other through economic, technological, social, and cultural cooperation, emphasising on collective self-reliance.

The SAARC's constitution, though noble in its vision, however, to many did not leverage the future of the region in a world increasingly dominated by regional trading blocs (Dash 2008). The establishment of the SAARC is a bold step and this paved a way to create legitimate institutional mechanism of cooperation within the region. Though, SAARC's performance as of today can be questioned, its position as a forum for discussion in a very politically volatile region cannot be doubted. South Asia is a "huge contiguous land mass criss-crossed by mighty rivers, with a wealth of natural resources, a variety of climatic conditions, and a common history and heritage, as also shared languages, literatures and religions" (Dubey 2007) and there is every reason for the continued commitment to regional cooperation. From an environmental point of view, it is argued that cooperation in South Asia, in part, been aggravated by the political mistrust among the countries in the region (Jha 2004). Despite these circumstances, Jha observes that SAARC offers the possibility of setting in place, a new institution that can work towards solving and mitigating persistent environmental problems.

Within the ambit of the SAARC's structure, in the past, there have been successful attempts in creating regional agreements[5] and conventions.[6] Since countries are expanding nuclear energy programmes in the region, an argument is made for establishing a nuclear risk community. Experts have opined that the misplaced and exaggerated security concerns of India and Pakistan are the biggest hurdles which hinders SAARC's growth (Dubey 2007). He emphasises that the countries, particularly India and Pakistan, need to shift the focus from State security to human security. This is exactly what a nuclear risk community would imply—a regional approach to understand the nuclear energy risk, and mitigating it through institutions to protect the people and environment in the region as a whole.

Nevertheless, successive annual summits of the SAARC, laid a good foundation for understanding common concerns, became apparent from 2007 onwards. The theme of the 2007 meeting was 'regional connectivity' and in 2008 the theme was 'people centric cooperation'. These initiatives raised hopes. The last two summits, the 16th and 17th Summits of the Head of the States of the SAARC countries witnessed the environment of mistrust slowly making way for an atmosphere of cooperation. This sense of positivism was largely due to the improved bilateral relations among the SAARC countries mainly in the context of regional economic progress, global financial crisis, terrorism, and natural and industrial disasters—all of which requires effective cooperation. Now, there seems to be a sense of urgency within the SAARC process to achieve a stable, responsive and progressive SAARC community. The recently concluded 18th SAARC Summit in Kathmandu, Nepal in November 2014 emphasised the importance of regional cooperation and declared that, "The Heads of State or Government expressed their strong determination to deepen regional integration for peace, stability and prosperity in South Asia by intensifying cooperation, inter alia, in trade, investment, finance, energy, security, infrastructure, connectivity and culture; and implementing projects, programmes and activities in a prioritised, result-oriented and time-bound manner".

There are 16 areas of cooperation currently authorised under the SAARC. The 16 stated areas of cooperation are: agriculture and rural, biotechnology, culture, energy, environment, economy and trade, finance, funding mechanism, human resource development, poverty alleviation, people to people contact, security aspects, social development, science and technology, communications and tourism.

---

[5] The SAARC Agreements that are in place are; (1) Agreement for establishment of the SAARC Arbitration Council; (2) Final Agreement on Avoidance of Double Taxation; (3) Final Agreement on Customs Matters; (4) Agreement on establishing the SAARC food bank; (5) Agreement on the South Asian Free Trade Area (SAFTA); (6) Agreement on the Establishment of South Asian Regional Standards Organisation (SARSO).

[6] The SAARC Conventions that have been adopted are, (1) SAARC Convention on Combating and Prevention of Trafficking in Women and Children for Prostitution; (2) Convention on Promotion of Welfare of Children; (3) Convention on Mutual Assistance on Criminal Matters; (4) SAARC Convention on Narcotics Drugs; (5) SAARC Regional Convention on Suppression of Terrorism and Additional Protocol on Terrorism.

4.5.3 The SAARC and Energy Initiatives

Energy is one of the 16 mandated 'areas of cooperation' within the SAARC, though the same has not seen any substantial progress in the context of cooperation between the member nations. Considering the diversity of resources in the region, energy cooperation could have benefited the countries of the region (Dhungel 2005).

In 2000, the process of regional cooperation in energy sector began with the establishment of a Technical Committee on Energy. The SAARC Energy Centre was established in pursuant to the 13th SAARC Summit, organised by Bangladesh on 12–13 November 2005. The mandate of the Centre is "to promote development of energy resources, including hydropower; and energy trade in the region; to develop renewable and alternative energy resources; and promote energy efficiency and conservation in the region". A visible outcome of the 18th Summit in November 2014 is the 'SAARC Framework Agreement for Energy Cooperation (Electricity)'. The Agreement enables SAARC countries to cooperate in order to improve the power availability to the whole SAARC region. It is further expected this agreement would facilitate integrated operation of the regional power grid.

In respect of nuclear energy, expectedly there has not been any substantial reference or discussion. The report, "Integrated Energy Potential of South Asia: Vision 2020" authored by Obaidullah (2010), commissioned by the SAARC Energy Centre, suggests that in view of future increased energy demand, South Asia will have to explore all possible energy options, including nuclear. The report concludes by stating, "The choice largely depends on the choice of the respective countries; how they will address the situation, independently or collectively". Another reference on nuclear cooperation is from the consensus recommendation from the conference on "Interdependence between Energy Policy and Other Sectoral Policies" (SAARC Energy Newsletter 2008). It states, "considering nuclear energy as another source of clean power, regional cooperation can be considered to explore the possibilities of building a regional nuclear power plant".

It is highly unlikely, in view of the relationships, international legal requirements and political considerations, that there will be any substantive cooperation in setting up a regional cooperative nuclear power plant in the South Asia. However, in case of a nuclear accident, it is very likely that South Asia as a community will bear the brunt of the same. Keeping in view of this, the priority for the SAARC countries is to build a firm foundation for a responsible nuclear future.

4.5.4 The SAARC and Disaster Management Initiatives

The Indian Ocean Tsunami of 2004 proved a turning point within the SAARC community, giving impetus to the establishment of a comprehensive action plan and response system at the regional level on natural disasters. Towards this, a Special Session of the SAARC Environment Ministers was held at Male on 25 June 2005. The Ministers concluded the meeting by adopting the Male Declaration, which decided that an Expert Group of the member countries shall meet at Dhaka, Bangladesh to formulate a Comprehensive Framework on Early Warning, Disaster Management and Disaster Prevention.

The Expert Group meeting was held from 7–9 February 2006 in Dhaka and developed a comprehensive framework on Disaster Management in South Asia (SDMC 2008). The framework of the Disaster Management in South Asia was approved by the SAARC Council of Ministers on 30 July 2006 and brought into effect by the Fourteenth SAARC Summit in New Delhi from 3–4 April 2007. The Framework provides a platform for the SouthAsian countries to:

i. Establish and strengthen the regional disaster management system to reduce risks, and to improve response and recovery management at all levels;
ii. Identify and elaborate country and regional priorities for action;
iii. Share best practices and lessons learnt from disaster risk reduction efforts at national levels;
iv. Establish a regional system to develop and implement regional programmes and projects for early warning;
v. Establish a regional system of exchanging information on prevention, preparedness and management of natural disasters;
vi. Create a regional response mechanism dedicated to disaster preparedness, emergency relief and rehabilitation to ensure immediate response;
vii. Create a regional mechanism to facilitate monitoring and evaluation of achievements towards goals and strategies.

In parallel, while developing a comprehensive framework, the SAARC Disaster Management Centre (SDMC) was also set up in October 2006 at the premises of the National Institute of Disaster Management, New Delhi. The Centre has been mandated to deal with all natural and man-made disasters through capacity building, advocacy and awareness programmes for the regions populace. The SDMC has been actively involved in formulation of policy guideline and framework for the SAARC member countries on Disaster Risk Reduction (DRR) schemes, covering various thematic issues of disasters in the region[7] (SDMC 2011).

[7] The Road Maps developed by SDMC are,

1. Regional Cooperation on Application of Science and Technology for Disaster Risk Reduction and Management
2. Regional Cooperation on Coastal and Marine Risk Mitigation Plan for South Asia
3. Regional Cooperation on Climate Change Adaptation and Disaster Risk Reduction
4. SAARC Roadmap for Mainstreaming Disaster Risk Reduction in Development
5. Community Based Disaster Management in South Asia
6. Regional Cooperation on Earthquake Risk Management in South Asia
7. Regional Cooperation on Landslide Risk Management in South Asia
8. SAARC Regional Road Map on Urban Risk Management in South Asia: Invest Today for a Better Tomorrow
9. Regional Cooperation for Drought Risk Management in South Asia

A study of the current SDMC initiatives indicates that the emphasis has been on natural disasters, leaving industrial disasters largely out of the purview of the discussions. Occasionally there are oblique references to the problem of man-made disasters. However, there has not been any step towards addressing it; mainly owing to the fact that such man-made industrial disasters are considered remote in this region. The 34th Session of the Council of Ministers that was held in Maldives on 9 November 2011 also emphasised its priority on natural disasters. The meeting concluded by stating, "The signing of the SAARC Agreement on Rapid Response to Natural Disasters; … at this Summit is a welcome development that will serve to further strengthen regional cooperation in their respective areas". As a follow-up to implement the Agreement ion Rapid Response, the 18th Summit held during November 2014 in Kathmandu, Nepal, the declaration states "they directed the relevant bodies/mechanisms for effective implementation of SAARC Agreement on Rapid Response to Natural Disasters".

The growth of nuclear energy programmes, especially the expansion by India and Pakistan, and the new programmes by Bangladesh and Sri Lanka now expose the region to a shared transboundary risk concern (however, small it may be). No doubt nuclear technology today is far superior and many countries have incorporated the best safety features in their existing reactors; nevertheless, the countries have to prepare for an unexpected nuclear accident. To combat this, the official responsible for nuclear emergencies at the National Disaster Management Authority, India is of the view that proper emergency preparedness plans must be in place so that there is minimum avoidable loss of life, livelihood, property and impact on the environment (28 August 2012).

4.6 Conclusion

As India and Pakistan expand their nuclear energy programmes, Bangladesh and Sri Lanka have concrete plans for nuclear energy. This development in a geographically unique contiguous land mass with rich natural resource, and a common history and heritage, masks the shared common nuclear risk. An accident within any of the state boundary could have a transnational impact. The legal and institutional systems that would respond to transboundary emergencies are not yet functional. The IAEA Emergency Conventions which provide the mandate for the provision of immediate relief remain in a legal wrangle due to the reservations placed on the same by India and Pakistan.

The SAARC has initiated major programmes on natural disasters and has had institutional and procedural success with the creation of response mechanisms. In the case of nuclear disasters, South Asia is in a pre-Chernobyl era and there is no regional or bilateral mechanism to make the States accountable for an unfortunate transboundary nuclear accident. A huge population with a rich biodiversity and culture thus remains at risk. It is fully within the SAARC's Constitution—to create a nuclear energy risk community that will allow regional concern to be shared, discussed and a structure to put in place to respond to these challenges.

References

Alexandrov, S. A. (2006). The compulsory jurisdiction of the international court of justice: how compulsory is it?' *Chinese Journal of International Law*, *5*(1), 29–38. Retrieved November 22, 2012, from http://chinesejil.oxfordjournals.org/content/5/1/29.full.pdf+html.

Banerjee, S. (2011). Address at the IAEA ministerial conference on nuclear safety, transcript, *International Atomic Energy Agency*. Retrieved June 20, viewed March 27, 2012, from http://www-pub.iaea.org/MTCD/meetings/PDFplus/2011/cn200/plenary/p_d1_india.pdf.

Begum, Z. (2008). *Status of Nuclear Activities of Bangladesh Atomic Energy Commission*. Paper Presentation at 16th WIN-Global Conference, 26–30 May, Marseilles. Retrieved January 21, 2011, from http://www.wincanada.org/uploads/filemanager/pdf/WIN-Global_2008_Conference/Country_Reports/Bangladesh_Full_Paper.pdf.

Bekker, P. H. F. (2000). Aerial incident of 10 August 1999 (Pakistan v. India). *The American Journal of International Law, 94*(4), 707–713.

Bloom, D. E., & Rosenberg, L. (2011). *The Future of South Asia: Population Dynamics, Economic Prospects, and Regional Coherence*. The program on the global demography of aging working Paper No. 68, Harvard. Retrieved April 10, 2012, from http://www.hsph.harvard.edu/pgda/working.htm.

Bukhari, S. S. H., & Rehman, A.-U. (2011). Pakistan-China nuclear deal & international fictions. *Berkeley Journal of Social Sciences*, *1*(3), 1–14. Retrieved May 14, 2012, from http://berkeleyjournalofsocialsciences.com/March5.pdf.

Chidambaram, R. (2006). India and the international atomic energy agency. *Atoms for Peace: An International Journal*, *1*(2–3) 137–144.

DAE. (2005). *Cooperation Agreement Between India and USA on Peaceful Uses of Nuclear Energy*. Retrieved May 24, 2011, from http://www.dae.gov.in/indous.pdf.

Dash, K C. (1996). The Political economy of regional cooperation in south asia, Pacific Affairs, *69*(2), 185–209.

Dhungel, D. (2005). South Asian association for regional co-operation (SAARC): Prospects for development. *The Pakistan Development Review*, *43*(4), 933–941. Retrieved September 12, 2011, from http://www.pide.org.pk/pdf/PDR/2004/4/PartII/Dwarika%20Dhungel.pdf.

Dubey, M. (2007). SAARC and South Asian economic integration. *Economic and Political Weekly, 42*(14), 1238–1240.

Haque, M. M. (2009). *Challenges and Opportunities to Introduce the First Nuclear Power Plant in Bangladesh*. Paper presentation made at IAEA technical meeting, 27–30 October. http://www-pub.iaea.org/MTCD/publications/PDF/P1500_CD_Web/htm/pdf/topic2/2S01_MM%20Haque_PM.pdf.

IAEA. (2008). *Communication Dated 10 September 2008 Received from the Permanent Mission of Germany to the Agency Regarding a "Statement on Civil Nuclear Cooperation with India"*, INFCIRC/734 (Corrected), IAEA. Retrieved April, 27, 2012, from http://www.iaea.org/Publications/Documents/Infcircs/2008/infcirc734c.pdf.

IAEA. (2009a). *The Agreement Between the Government of India and the International Atomic Energy Agency for the Application of Safeguards to Civilian Nuclear Facilities*, Information Circular INFCIRC/754. IAEA. Retrieved May, 14 2011, from http://www.iaea.org/Publications/Documents/Infcircs/2009/infcirc754.pdf.

IAEA. (2009b). *Energy, Electricity and Nuclear Power Estimates for the Period up to 2030*. IAEA Reference Data Series No. 1, IAEA. Retrieved June, 22 2012, from http://www-pub.iaea.org/MTCD/publications/PDF/RDS1-29_web.pdf.

IAEA. (2009c). *IAEA Revises Nuclear Power Projections Upward*. Retrieved July 10, 2012, from http://www.iaea.org/newscenter/news/2009/npprojections.html.

IAEA. (2010). *Agreement Between the Government of India and the International Atomic Energy Agency for the Application of Safeguards to Civilian Nuclear Facilities: Addition to the List of Facilities Subject to Safeguards Under the Agreement*. INFCIRC/754/Add.2, IAEA. Retrieved April 22, 2012.

IAEA. (2011a). *IAEA Projects Slower Nuclear Growth After Fukushima*. Retrieved July 24, 2012, from http://www.iaea.org/newscenter/news/2011/nuclgrowth.html.

IAEA. (2011b). *Bangladesh Progresses Toward Nuclear Power*. Retrieved April 22, 2012, from http://www.iaea.org/newscenter/news/2011/bangladeshprog.html.

IAEA. (2014). *Power Reactor Information System-Pakistan*. Retrieved August 20, 2014, from htt p://www.iaea.org/pris/CountryStatistics/CountryDetails.aspx?current=PK.

ICJ. (2000). Aerial Incident of 10 August 1999 (Pakistan v. India), Jurisdiction of the Court, I. C. J. Reports, p. 12

Jha, U. C. (2004). Environmental issues and SAARC. *Economic and Political Weekly, 39*(17), 1666–1671.

Ministry of Environment, Science and Technology, Nepal 2007, *National Nuclear Policy 2007*. Retrieved December 24, 2011, from http://www.most.gov.np/np/images/stories/policies/nationalnuclearpolicy-2007.pdf.

Nuclear Non-Proliferation Act. (1978), 22 U.S.C. § 3201.

Obaidullah, A. N. M. (2010). *Integrated energy potential of south Asia: vision 2020*. Islamabad: SAARC Energy Centre Project Report.

Obino, F. (2009). The political challenge for South Asia and beyond. *Economic and Political Weekly, 44*(9), 118–125.

Pakistan Nuclear Regulatory Authority. (2008). Regulations on management of a nuclear or radiological emergency—(PAK/914) (Rev.0). *The Gazette of Pakistan*. Retrieved August 24, 2011, from http://www.pnra.org/legal_basis/PAK-914.pdf.

Parvez, A. (2011). Address at the 55th IAEA General Conference, transcript, *Pakistan Atomic Energy Commission*. Retrieved September 19–23, viewed July 07, 2012, from http://www.iaea.org/About/Policy/GC/GC55/Statements/pakistan.pdf.

Press Trust of India. (2012). *Pakistan Plans to Build Two Nuclear Power Plants in Sindh, NDTV*. Retrieved April 23, viewed June 5, 2012, from http://www.ndtv.com/article/world/pakistan-plans-to-build-two-nuclear-power-plants-in-sindh-201022.

Rahman, K.M.R. (2012). *Bangladesh experience in self-evaluation of nuclear infrastructure*. Paper Presentation at the Iaea Technical Meeting on Evaluation Methodology of the Status of National Nuclear Infrastructure Development and Integrated Nuclear Infrastructure Review. Retrieved June 18–20, viewed May 17, 2012, from http://www.iaea.org/NuclearPower/Downloads/Infrastructure/meetings/2012-06-18-20-TM-Vienna/13.Bangladesh_Experience_in_Self_Evaluation.pdf.

Rajya Sabha. (2012a). Unstarred Question no. 4499, transcript, *Department of Atomic Energy*. Retrieved May 17, viewed August 22, 2012, from http://www.dae.nic.in/writereaddata/rsus4499.pdf.

Rajya Sabha. (2012b). Global Centre for Nuclear Energy Partnership, Unstarred Question no. 3724, transcript, *Department of Atomic Energy*. Retrieved May 10, viewed August 22, 2012, from http://dae.nic.in/writereaddata/rsus3724.pdf.

Rajya Sabha. (2012c) *Sri Lankan Concern Over Kudankulam Plant, Unstarred Question NO. 2950*. Retrieved May 03, viewed August 22, 2012, from http://dae.nic.in/writereaddata/parl/rsus2950.pdf.

Ranawaka, C. (2012). Address at the 54th regular session of the IAEA general council, transcript, *International Atomic Energy Agency*. Retrieved September 20, viewed November 25, 2012, from http://www.iaea.org/About/Policy/GC/GC54/Statements/srilanka.pdf.

Rasel, R. A. (2014). Plans afoot for another nuke power plant with Japanese assistance, *Dhaka Tribune*. Retrieved June 15, viewed September 12, 2014, from http://www.dhakatribune.com/foreign-affairs/2014/jun/15/plans-afoot-another-nuke-power-plant-japanese-assistance#sthash.M7GtMHiG.dpuf.

SAARC Disaster Management Centre (SDMC). (2008). *Comprehensive framework on disaster management*. New Delhi: SDMC.

SAARC Disaster Management Centre (SDMC). (2011). *SAARC road maps on risk management in South Asia*. New Delhi: SDMC.

Schwartz, A. J. (2006). International Nuclear Third Party Liability Law: The Response to Chernobyl in the *International Nuclear Law in the Post-Chernobyl Period*, OECD, Paris

Scott, G. L., & Carr, C. L. (1987). The ICJ and compulsory jurisdiction: The case for closing the clause. *The American Journal of International Law, 81*(1), 57–76.

Shah, S. (2010). *Pakistan pushes US for nuclear technology deal*. The Guardian, 22 March. Retrieved June 27, 2011, from http://www.guardian.co.uk/world/2010/mar/22/pakistan-us-nuclear-technology-deal

Siddiqui, Z. H., & Qureshi, I. H. (2005). Nuclear power in Pakistan. *The Nucleus, 42*(1–2), 63–66.

Syed, B. S. (2014). 8,900 MW nuclear power generation planned, *The Dawn 02 January*. Retrieved August 10, 2014, Published Jan 02, 2014 06:43 am, from http://www.dawn.com/news/1077816.

The Daily Times. (2013). *Nawaz Launches Country's Largest Nuclear Power Project: Pakistan to Make Nuclear Largest Energy Source*. Retrieved November 27, viewed August 1, 2014, from http://www.dailytimes.com.pk/islamabad/27-Nov-2013/nawaz-launches-country-s-largest-nuclear-power-project-pakistan-to-make-nuclear-largest-energy.

The South Asian Association for Regional Cooperation (SAARC). (2008). Report on SAARC conference on interdependence between energy policy and other sectoral policies. *SAARC Energy Newsletter, 2*(3), 1–2.

United Nations Environment Programme (UNEP). (2009). *South Asia Environment Outlook 2009*. Retrieved April 16, 2011, from http://www.saarc-sec.org/userfiles/SAEO%202009.pdf.

UN Population Statistics (2011). *Population and Vital Statistics Report Statistical Papers Series* A Vol. LXIII, Retrieved 15 April 2014, http://unstats.un.org/unsd/demographic/products/vitstats/Sets/Series_A_2011.pdf

World Nuclear Association. (2014). *Nuclear Power in Pakistan*. Retrieved August 05, 2014, from http://www.world-nuclear.org/info/Country-Profiles/Countries-O-S/Pakistan/.

Chapter 5
South Asian Nuclear Risk Zone: A Mapping Exercise

Abstract In South Asia, as India and Pakistan rapidly expand their nuclear power programme, Bangladesh has started construction of a nuclear power plant and Sri Lanka has shown interest. In order to understand the implications of a transboundary nuclear accident in the SAARC region, this chapter using Geographical Information Systems plots risks zones to visually represent the emergency planning zones. The planning zones taken for the study are 30, 50 and 100 km as emergency zones; and 300, 500 and 1000 km as ingestion and conservative risk zones. These zones are based on the international criteria mandated by IAEA, the USA, France, the UK, India, and experiences of impact from Chernobyl and Fukushima. The mapping result shows existence of transboundary risk in South Asia. This shared risk call for SAARC countries to address the expanding nuclear energy programme as a shared concern.

Keywords South Asian nuclear accident risk zone • Nuclear risk assessment and transboundary impacts • Classification of nuclear risk zones • Mapping on nuclear risk in South Asia/SAARC

5.1 Introduction

The nuclear energy development discourse, from its inception, has constantly dealt with the issue of risk, including the perception of risk. Nuclear technology and its development have always been perceived as risky (Green 1973; Kasperson et al. 1980). The outcomes of the Hiroshima and Nagasaki bombings, accident at Chernobyl, and the 2011 Fukushima nuclear power plant accident, have all contributed to the continuing fear of the use of nuclear energy. One may attribute this fear of nuclear energy to the lost years of openness, on account of the secretive nature of the nuclear energy programmes. In most cases the combined military

M.P. Ram Mohan, *Nuclear Energy and Liability in South Asia*,
DOI 10.1007/978-81-322-2343-6_5

and civilian nuclear programmes were pursued with little regard for whether people knew of the benefits and the risk associated with such technologies, and have contributed much to the 'unknown risk' perception. Communication plays a major role in how perceptions are moulded and remain in the public memory.

Many countries in Europe, Asia and Africa have decided to permanently discard their nuclear weapon programmes with the aim of pursuing nuclear technology solely for peaceful purposes (energy, medical and other applications). This shift from military application to civilian use has provided a sense of openness in public engagement, which in many cases transpired into the public engaging and debating on the use of nuclear energy as part of the energy mix. However, this has not been the case in South Asia, particularly with India and Pakistan. It would be interesting to witness and study how Bangladesh and Sri Lanka, with no nuclear weapons programmes take their infant civilian nuclear energy programmes forward.

This chapter undertakes a technical study on the nature of the transboundary and regional risk that Nuclear Power Plants (NPPs) pose in South Asia. Nuclear risk zones are identified using the Geographical Information System (GIS). The study concerns itself solely with civilian NPPs. Ancillary civilian nuclear facilities[1] and military facilities[2] are not included in the maps, description or analysis.

5.2 Technical Study: Structuring a Nuclear Risk Zone Within the SAARC

An NPP site usually contains more than one reactor, and in case of a nuclear accident occurring in one reactor, it may or may not lead to an accident in the other reactors. For the purpose of this study, a transboundary nuclear accident refers to an accident in one country, which has or is likely to have a detrimental impact

[1] Other nuclear facilities include, uranium mining, fuel conservation, enrichment, fuel fabrication, storage of spent fuel, reprocessing of spent fuel, waste management, waste storage and transportation of nuclear materials, etc. All these activities are integral to the operation of nuclear power plant, however, their impact varies.

[2] The caveat being that in both Pakistan and India, the nuclear power programme may in one way or the other be connected to military applications. Be as the case may the purpose of the research is not to dwell into what constitutes a civilian application or military one. Power Plant that is generating power and is connected to grid is considered a civilian power plant. In IAEA parlance all NPPs that are under the IAEA safeguards are considered civilian programmes. The problem for the application of this description in the South Asian context is that, though in the recent years many NPPs were placed under safeguards by both India and Pakistan, both the countries even today remains outside the NPT regime. An unnecessary controversy is sought to be avoided in this research. Expansion of nuclear power programmes within SAARC is a shared concern in respect to probable transboundary consequences.

in another country or region. The gravity of the adverse impact is based on the 'zones'. The zones used for this study were based on the criteria identified by,

(a) *The International Atomic Energy Agency (IAEA), Arrangements for Preparedness for a Nuclear or Radiological Emergency (Safety Guide No. GS-G-2.1, 2007)*
(b) Country requirements as mandated by national authorities—US, France, UK and India, and
(c) The scale of the impact observed in Chernobyl and Fukushima.

These demarcated 'zones' refer to areas for immediate evacuation, emergency evacuation, evacuation based on need and larger impact zones.

Almost all the NPPs that are in operation in South Asia are in the range of <100–1000 MW. Many new projects, especially that are been pursued through international collaboration consists mostly of high-capacity >1000 MW[3] reactors. The IAEA risk zones provide for differentiated zoning based on whether the plant is less than 1000 MW or more. Further, in respect of the siting of the NPPs, as informed by a senior official at the Homi Bhaba National Institute (11 July 2012) through email communication, that there are no legal prescription or limitation as of now, nationally or internationally in the siting and constructing the NPPs close to international borders (land or sea). However, in the post Fukushima environment, European Union is establishing cross-border consultative process with respect to sitting of nuclear power projects. In the absence of any such norms within SAARC, it appears that the possible transboundary impact, in case of an accident, has not been taken into consideration within South Asia.

The technical study uses GIS software (ESRI's ArcGIS Version 10) to plot risk zones. The locations of NPP were obtained from government data and reported news. The coordinates (lat and log) were identified using Google maps.

5.2.1 Classification of Risk and Risk Zones

It is acknowledged that even after adequate precautions being taken in the design, operation and conduct of nuclear activities, there remains a possibility, however remote, of a failure leading to radiological emergency. This emergency could arise from either an accidental or an intentional act, the impact of which may be limited to within the facility or affect the population at large. The objective of the IAEA

[3] The current nuclear power generating countries (India and Pakistan) mostly operate low-capacity reactors of 100, 220 and 500 MW and in few cases in India around 700 MW (if taken as single reactor per site). Only KNPP which became critical in 2014 have a capacity of 1000 MW. Nearly, all new projects being pursued by Bangladesh, India, Pakistan and Sri Lanka (possibly) are being undertaken with international collaboration will have high-capacity reactors, i.e. 1000 MW or more. In India, new green field site will have total capacity of more than 6000 MW. *For more, see* Chap. 4 *and* Table 4.3.

Safety Requirements on Preparedness and Response for a Nuclear or Radiological Emergency (*Safety Requirements Series No. GS-R-2, 2002*) highlight the necessity of safety requirements, and states:

> 1.5. This Safety Requirements publication establishes the requirements for an adequate level of preparedness and response for a nuclear or radiological emergency in any State. Their implementation is intended to minimise the consequences for people, property and the environment of any nuclear or radiological emergency.
> 1.6. The fulfilment of these requirements will also contribute to the harmonisation of arrangements in the event of a transnational emergency.
> 1.7. These requirements are intended to be applied by authorities at the national level by means of adopting legislation, establishing regulations and assigning responsibilities.

Based on the level of risk, the IAEA (*Safety Requirements Series No GS-R-2, 2002*), categorises nuclear radiation threats into Category I, II, III, IV and V. Threat Category (TC) I is from the NPPs and TC II is from research reactors. Emergencies from both TC I and II are considered severe, resulting in a major radioactive release. TC III and IV emanate from industrial applications or unauthorised/illegal activities.[4] TC V is perhaps the most important category for this study, results in significant contamination of food, water, etc., even at places far away from the actual accident site. TC V is conditional on the occurrence of a significant release of radioactive material from a facility in TC I or II.

Further, in order to have a disaster mitigation response from these Threats, the *IAEA Arrangements for Preparedness for a Nuclear or Radiological Emergency* (*Safety Guide No. GS-G-2.1, 2007*) specifies the demarcation of two emergency risk zones based on the capacity of the NPP. The zones are, the precautionary action zone (PAZ)[5] and the urgent protective action planning zone (UPZ).[6] For TC V, the guidelines prescribe that adequate contingency requirements must be taken care of, according to the intensity of the threat from category I and II.

Currently, the IAEA (2007) specifies that reactors with 1000 MW and higher capacity, require 3–5 km as PAZ, and 5–30 km as UPZ. For reactors with capacity between 100 and 1000 MW, the PAZ is 0.5–3 km and the UPZ is the same as in the case of 1000 MW and higher capacity reactors (i.e. 5–30 km).

In case of substantial release of radiation, the PAZ represents an area where adequate arrangements should be made for protection, aimed at preventing or reducing severe detrimental effects (IAEA 2007). A wide range of protective activities can be undertaken at the time of an accident, including but not limited to

4 For the purpose of this study, TC II, III and IV are not taken into consideration.

5 'A PAZ, for facilities in TC I, for which arrangements shall be made with the goal of taking precautionary urgent protective action, before a release of radioactive material occurs or shortly after a release of radioactive material begins, on the basis of conditions at the facility (such as the emergency classification) in order to reduce substantially the risk of severe deterministic health effects' (IAEA 2007, p. 14).

6 'An urgent protective action planning zone, for facilities in TC I or II, for which arrangements shall be made for urgent protective action to be taken promptly, in order to avert dose off the site in accordance with international standards' (IAEA 2007, p. 14, 15).

distribution of iodine tablets, requiring persons to stay indoors, distribution of non-contaminated food and supplies, and evacuation in severe cases.

Thus, the size of NPP (whether more or less than 1000 MW) only affects the PAZ. The PAZ covers the first line of impact in a nuclear accident. For the purpose of this study, as there are no existing or proposed NPPs within a distance of 5 km from the international border, therefore, the PAZ has not been separately demarcated on the maps.

In respect of ingestion (food restriction) zones, the present requirements—*Safety Requirements Series No. GS-R-2, 2002*, does not specify that a 'food restriction' planning distance is required to be established. They only require that, within the areas where the activities in TC V are conducted, 'effective agricultural countermeasures' shall be taken. On the other hand, the *IAEA Method for Developing Arrangements for Response to a Nuclear or Radiological Emergency* (EPR-Method 2003)[7] specifies the 'food restriction planning radiuses' as 300 km for reactors upwards of 1000 MW and 250–300 km for reactors with 100–1000 MW capacity. The use of 'food restriction' planning distance described in the EPR-Method is a practical approach that countries may consider as a way for meeting the requirements relating to TC V.

Interestingly, the present requirements or the safety guide that is been followed does not recommend a specific 'food restriction' planning distance. It recommends arrangements (i.e. issuing instructions to protect food and water supplies and to restrict the use of drinking water) that need to be planned and made in advance for areas where the activities in TC V are conducted.

Subsequent to the Fukushima accident, Delves (4 October 2012) from the IAEA vide email stated that, these specifications are under review for possible publication in 2014–2015. He further informed that, 'the expected draft IAEA Safety Requirements are likely to include requirements for a "food restriction" planning distance, as described in EPR-Method'. According to Delves, 'the relevant distance may remain at 300 km for NPPs of 250 MW and above. The relevant requirements may be made more explicit in terms of requirements for (a) moving grazing animals onto stored (covered) feed, (b) protecting drinking water supplies that derive from rainwater, and (c) restricting consumption and distribution of local produce, milk from grazing animals, water from rainwater and commodities until further assessments have been performed'. In such a case, irrespective of reactor capacity this study takes 300 km as 'food restriction' planning distance.

The IAEA Safety Standards are only recommendatory in nature and do not impose any legally binding obligations on the countries. They reflect best practices; several countries have adopted more stringent demarcation zones, as detailed

[7] The Emergency Preparedness and Response (EPR) document is an update of IAEA-TECDOC-953. The aim is to: '(1) fulfil in part the IAEA's function under Article 5.a(ii) of the Assistance Convention, and (2) to provide a compendium of best practice for planners aiming both to comply with the Requirements [2] and to improve their own capabilities for responding to radiation emergencies, while the Secretariat facilitates consensus on formal guidance for meeting the Safety Requirements' (IAEA 2003, p. 1).

subsequently. In the case of both Chernobyl and Fukushima, demarcation of affected area subsequent to the accident is based on true impact of radiation exposure rather than zoning requirements prescribed by the IAEA Safety Standards.

5.2.2 Risk Zones and the United States

The national mandates of risk zones vary from country to country. The US Nuclear Regulatory Commission (USNRC) specifies (Code of Federal Regulations 1993) in general terms, the 'plume exposure pathway',[8] the emergency protection zone for NPPs is 10 miles (~16 km). The 'ingestion pathway' emergency protection zone[9] is 50 miles (~80 km). Further, the regulations make it clear, that the exact size and configuration of these zones is not sacrosanct (Code of Federal Regulations 1993; Federal Emergency Management Authority 2002), but will be determined according to particular accident situations and local conditions, such as demography, topography, land characteristics, access routes and jurisdictional boundaries.

5.2.3 Risk Zones and India

In India, risk zones are demarcated by the AERB guideline which lists three zones for emergency management (AERB Safety Study 2000, 2005, 2009; Roshan et al. (n.d)). They are the exclusion zone, the sterilised zone and the emergency planning zone (EPZ). The exclusion zone covers 1.5 km around the NPP which is earmarked as a no activity area. The sterilised zone covers an area of 5 km radius where natural growth of population is permitted and planned expansion of activities leading to enhanced population is regulated. And the EPZ—a zone up to 16 km radius from the plant—provides a basic geographic framework on implementing measures as part of a graded response in the event of an emergency.

An emergency is considered off the site[10] (off the site of the NPP) if it is in or beyond the area of the EPZ, referred as public domain area. Any emergency activity beyond 16 km is the responsibility of the local government and district

[8] 'Plume Exposure Pathway is declared as an area within a 10-mile radius surrounding a nuclear station. The principal exposure sources from this pathway are: (1) whole body exposure to gamma radiation from the plume and deposited material, and (2) inhalation exposure from the passing radioactive plume' (US NRC 2009).

[9] 'Ingestion Exposure Pathway is an area within approximately a 50-mile radius surrounding a nuclear station. The principal exposure from this pathway would be from the ingestion of contaminated water or food' (US NRC 2009).

[10] 'Accident Condition/Emergency Situation involving excessive release of radioactive materials/hazardous chemicals from the plant to the public domain calling for intervention' (AERB 2007).

administration. A comparative review of the IAEA, the US and India can be as seen in Table 5.1. The researcher could not access information on emergency risk zones in Pakistan.

5.2.4 Risk Zones and France

France identifies two risk zones surrounding each NPP. An emergency planning zone of 5 km radius and a second emergency planning zone with a 10 km radius. In the case of the 5 km radius zone, a detailed evacuation plan is required to be in place. Within the 10 km radius, sheltering is pre-planned and iodine tablets have to be distributed. More importantly, France recognises that protective measures may need to be extended beyond the 10 km radius, in case the conditions so warrant (Kubayani and Kirchsteiger 2005).

5.2.5 Risk Zones and The United Kingdom

In the UK, the Radiation (Emergency Preparedness and Public Information) Regulations 2001 (REPPIR) requires each nuclear site to have a 'detailed emergency planning zone' (DEPZ or recently referred REPPIR off-site emergency planning area). Emergency planning areas are designated within each nuclear site, where there is a potential off-site release of radioactivity. Accordingly local authorities are required to prepare emergency plans in case of an off-site emergency. Office of Nuclear Regulator (ONR), established as a statutory Public Corporation on 1 April 2014 under the Energy Act 2013, determines the area based on two principles:

- A technical assessment of the area likely to be affected by a radiation emergency as defined in REPPIR;
- An assessment of the practical and strategic implications of implementing countermeasures and aid those members of the public who are likely to be affected by a radiation emergency. This assessment involves the consultation with local authorities and includes local demographic and geographical considerations.

The boundary of the zone ranges from 1 to 5 km. The regulation also recognises that radiation emergencies could occur beyond the DEPZ. For the same, there is a requirement to incorporate (extendibility) emergency plans beyond the DEPZ. After the nuclear accident in Fukushima, ONR revised the principles to determine a nuclear site's DEPZ. It states, 'from now on, in consultation with local authorities, ONR will base its decisions on consideration of practical and strategic factors, such as local demographic and geographical features that influence where the boundary of the DEPZ lies, as well as technical assessments'. According to them, the purpose of this revision is to enhance preparedness, minimise response times, and, ultimately, improve protection of the public, and will be applied to nuclear sites on a rolling basis when their DEPZs are reassessed every 3 years under the Radiation (Emergency Preparedness and Public Information) Regulations 2001 (ONR 2015).

5.2.6 Chernobyl and Fukushima: Extent of Impact

As per the IAEA's Frequently Asked Questions on Chernobyl, the radioactive fallout from Chernobyl extended up to 150,000 km^2, covering the whole of Belarus, Ukraine and parts of Russia. The radioactive footprint of the disaster stretched over 500 km. Even today, an area up to 30 km from the NPP is uninhabitable, and has been declared as an 'exclusion zone'. *The Chernobyl Forum Report*[11] (IAEA 2006a, b) states that the radiation release continued for close to 10 days, contaminating large parts of Europe. The United Nations Scientific Committee on the Effects of Atomic Radiation (UNSCEAR 2000) and the International Chernobyl Project (IAEA 1991; Chernobyl Forum 2006) reports suggest that through wind and storm, the radiation travelled to most of Europe, including Sweden, Finland, Austria, Norway, Bulgaria, Switzerland, Greece, Slovenia, Italy and Republic of Moldova. The UNSCEAR (2008) and the UNEP/GRID-Arendal (Rekacewicz 2007) radiation impact map suggests that the deposition of radioactive particles was even found as far as in Greece, Turkey, the UK and Ireland. Independent private funded researches like the Greenpeace (2006) and the TORCH Report (Fairlie and Sumner 2006) argue that the geographical scale of the accident is much larger than suggested by the government funded and sponsored reports. Whatever the case may be, it can be reasonably concluded that the damage was not limited to Ukraine, Russia and Belarus, but extended to a larger part of Europe with varying intensity.

In the case of Fukushima, initially the Japanese Government conducted emergency evacuation in a radius of 20 km from the NPP, and declared the area as 'no-entry zone'. Subsequently, when the results of the accident showed much greater impact, the emergency zone was extended to 30 km and designated as 'evacuation zone in case of emergency'(Prime Minister Office Press Briefing 2011). Further, several villages within a 30–50 km radius, north of the plant, were declared as a 'planned evacuation zone' (Sugimoto 2012). To contain probable adverse effects,

[11] The Chernobyl Forum consists of total of eight UN specialised agencies. They are IAEA, World Health Organization (WHO), United Nations Development Programme (UNDP), Food and Agriculture Organization (FAO), United Nations Environment Programme (UNEP), United Nations Office for the Coordination of Humanitarian Affairs (UN-OCHA), United Nations Scientific Committee on the Effects of Atomic Radiation (UNSCEAR), and the World Bank. The governments of Belarus, the Russian Federation and Ukraine were part of the study.

even in Tokyo which is almost 250 km away from Fukushima, the government advised the citizens not to give tap water to infants. Agricultural areas around Fukushima, consisting of 8 prefectures, were considered contaminated, and consumption of agricultural products from these areas was banned. According to a UNSCEAR (2012) press release, the full impact is difficult to predict, and its satisfactory analysis may take years (see also Dauer 2011). A US Congressional Report (Buck et al. 2012) states that, 'radioactive contaminants from Fukushima appear to have become sufficiently dispersed over the time that they will not prove to be a serious health threat elsewhere, unless they bio-accumulate in migratory fish or find their way directly to another part of the world through food or other commercial products'.

Interestingly, the advisory of the foreign governments to their nationals in Japan has also been mixed. The US advised its citizens to leave the areas within approximately 80 km from the plant; South Korea advised its citizens to evacuate area within and beyond 80 km; Spain advised 120 km; Germany advised its citizens to leave even the metropolitan area of Tokyo which is more than 250 kms from Fukushima (Stellfox *I-Nuclear* 2012 and The Korea Herald, 3 March 2011).

Irrespective of the demarcation zones specified, if we extrapolate the data from the events of Chernobyl and Fukushima, though many countries responded on evacuation based on perceived risk and fear, a case, however, can be made out for a larger area mitigation and management response system.

Table 5.1 provides an overview of the comparative demarcation zones employed internationally.

Table 5.1 Comparative demarcation of risk zones

Category	km
IAEA	
Precautionary Action Zone (PAZ)	0.5–3
Urgent Protective Action Planning Zone (UPZ)	5–30
Food Restriction Zone	300
USA	
Plume Exposure Pathway Emergency Protection Zone	16
Ingestion Pathway Emergency Protection Zone	80
France	
First Emergency Plan Zone	5
Second Emergency Plan Zone	10
United Kingdom	
Detailed Emergency Plan Zone (DEPZ)	1–5
India	
Exclusion zone	1.5
Sterilised zone	5
Emergency Planning Zone (EPZ)	16

Source Compiled by Researcher

5.3 Plotting the Probable Impact Scenarios in South Asia

South Asia is a geographically connected region, with only the Maldives being separated from the continent (Map 5.1). Sri Lanka, though not geographically connected, is only 50 km away from India and 250 km from the Kudankulam, the Indian NPP. The intent of mapping the NPP location and plotting the risk zones is to bring out the likely transboundary impact in case of an accident. Since, multiple reactors are clustered around a single location, each site and not individual NPPs has been mapped using GIS.

The risk zone radius colour coding[12] is, Red—30 km; Orange—50 km; Blue—100 km; Green—300 km, Yellow—500 km and dotted line—1000 km. The reason for taking 30, 50 and 100 km are based on the IAEA requirements; changes now sought by countries after the Fukushima accident; and also based on the general advisory level given by different governments to its citizens during the Fukushima accident (the advisory may be very well based on immediate reaction, still that shows the risk that is perceived by nations). The 300 km zone is included based on the IAEA EPR-Method 2003 and, subsequent to Fukushima IAEA indicated that 300 km will likely be an ingestion zone in the revised safety requirements (personal communication with IAEA). 500 and 1000 km risk zones were plotted based on the significant impacts that were witnessed subsequent to the Chernobyl and Fukushima (impact on sea). The GIS exercise is based on a location specific modelling of risk zone radius within SAARC and does not consider wind and water movements.

5.3.1 Operational NPPs in South Asia

In South Asia, as discussed earlier, only India and Pakistan are currently pursuing nuclear energy programmes. Map 5.1 plots the currently operating NPPs in India and Pakistan.

In India, the operating NPPs are, Tarapur in Maharashtra with two 160 and two 540 MW reactors; Rawatbhata in Rajasthan consisting of a 100 MW, a 200 MW, and four reactors with the capacity of 220 MW each; Kalpakkam in Tamil Nadu with two 220 MW reactors; Narora and Kakrapar in Uttar Pradesh with two reactors—each having 220 MW capacity; Kaiga, Karnataka with four reactors, each with the capacity of 220 MW and Kudankulam, Tamil Nadu currently with one reactor with the capacity of 1000 MW and another is expected to be in operation during 2015–2016. In Pakistan, the Karachi NPP has a capacity of 125 MW and the Chashma NPP in Punjab Province consists of two reactors, each with a capacity of 300 MW.

Even though both the countries have a long history of nuclear power, the total contribution of nuclear power to the overall national energy scenario is marginal.

[12] Red, Orange and Blue zones are sometimes referred in this chapter as conservative estimates.

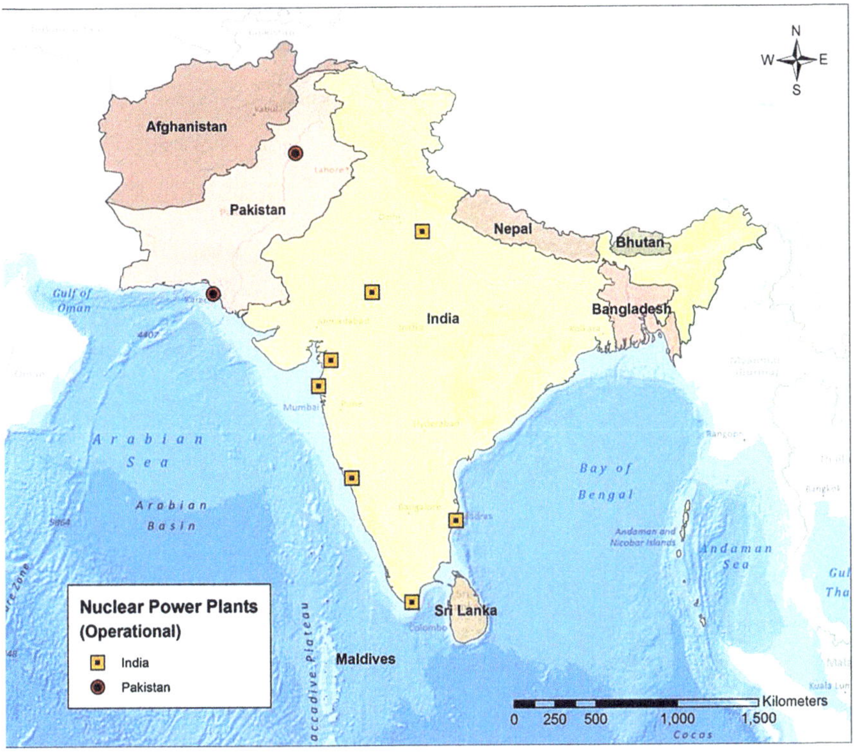

Map 5.1 Operational NPPs in South Asia. *Source* Results of technical study

The primary reason for this low output, as explained in Chap. 4, is that both these countries are non-members of the international nuclear disarmament treaty—the NPT. Therefore, the progress of the programmes has been slow, as the technical collaboration with other counties was prohibited. Facilitated by the US, India became a part of world nuclear community, when in 2008 IAEA and NSG agreed for India-specific exemptions.

Within SAARC, one of the first intergovernmental discussions on transboundary impacts of the NPPs was between India and Sri Lanka, in respect to Kudankulam NPP in Tamil Nadu. The plant which was commissioned in 2014 is located around 250 km aerially from Sri Lanka. Again as discussed in Chap. 4 Sri Lanka did raise their concerns on the possible transboundary consequences.

Other than KNPP 1, most of the NPPs in operation are small capacity reactors, and are also not in the national border areas. On account of this, it can be argued that the risks of a major transboundary consequence under the IAEA PAZ and UAZ from the existing reactors are low (Map 5.2). However, in the case of KKNPP an accident impact in the 100 km zone may extend to the territorial waters between India and Sri Lanka. Here an accident may not have immediate human impact, but may have a bearing on the flora and fauna and consequent livelihood of people across the boundary.

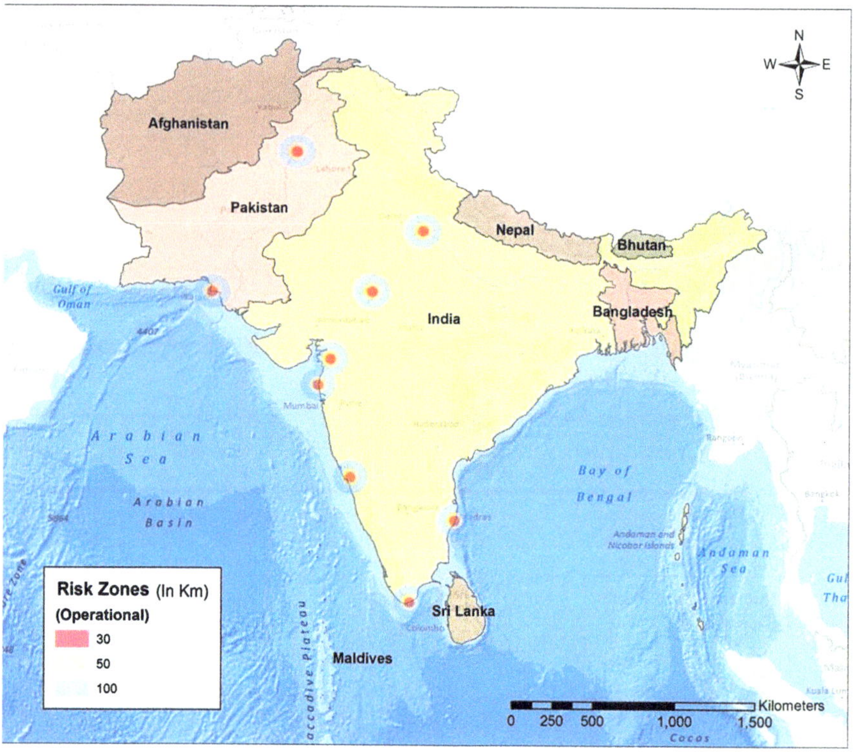

Map 5.2 Risk zones—operational NPPs. *Source* Results of technical study

The results from the extended risk zone map plotted on Map 5.3 show that there may not be any substantial impact. However, there remains a risk of transnational ingestion through contaminated water, air, food, etc. The results for 300 and 500 km zone from the operational reactors provide a visual of the scope of the impact likely in Nepal, Pakistan and Sri Lanka due to the NPPs in India, and on Afghanistan and India due to the NPPs in Pakistan. Depending on the extent of the ingestion risk, it may require transboundary mitigation in these countries. In other cases of the under-construction plants, the transboundary risk appears minimal—varies between remote probability to 'no risk'.

5.3.2 NPPs Under Construction

India and Pakistan, in addition to the currently operational power plants, are also undertaking the construction of new NPPs. Bangladesh has recently initiated steps to build a 1000 MW reactor in Rooppur (Map 5.4). The plants under construction

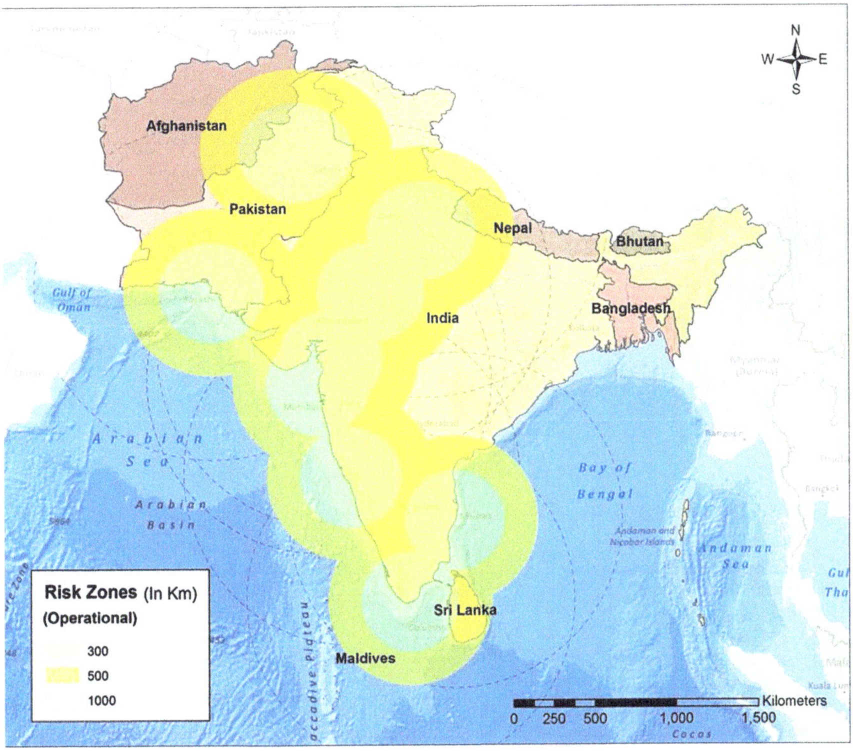

Map 5.3 Extended risk zones—operational NPPs. *Source* Results of technical study

in India are Rawatbhata, Rajasthan having two reactors of 700 MW each; one plant of 500 MW Kalpakkam and one plant of 1000 MW in Kudankulam, Tamil Nadu; and Kakrapar, Gujarat with two plants of 700 MW each. In the case of Pakistan, two plants with the capacity of 300 MW each are under construction at Chashma in Punjab Province.

Almost all the under-construction projects in India (except Kudankulam) and Pakistan are planned in the same location as of the older plants. Rooppur in Bangladesh and Kudankulam in India are the green field sites.

Rooppur is close to the state of West Bengal, India, and is only 50 km aerially from the international border with India. Under the conservative gradation of emergency zones (Map 5.5), the nuclear energy programmes that are being planned in Bangladesh and India would positively require a bilateral risk assessment and mitigation approaches. In the case of Rooppur, the zone of transboundary risk covers India. For Kudankulam-II as explained above (Map 5.2), the risk extends to the territorial waters between India and Sri Lanka.

If we consider extended risk zones for the under-construction NPPs, the likelihood of ingestion exposure is substantially higher (Map 5.6). The plotting of 300

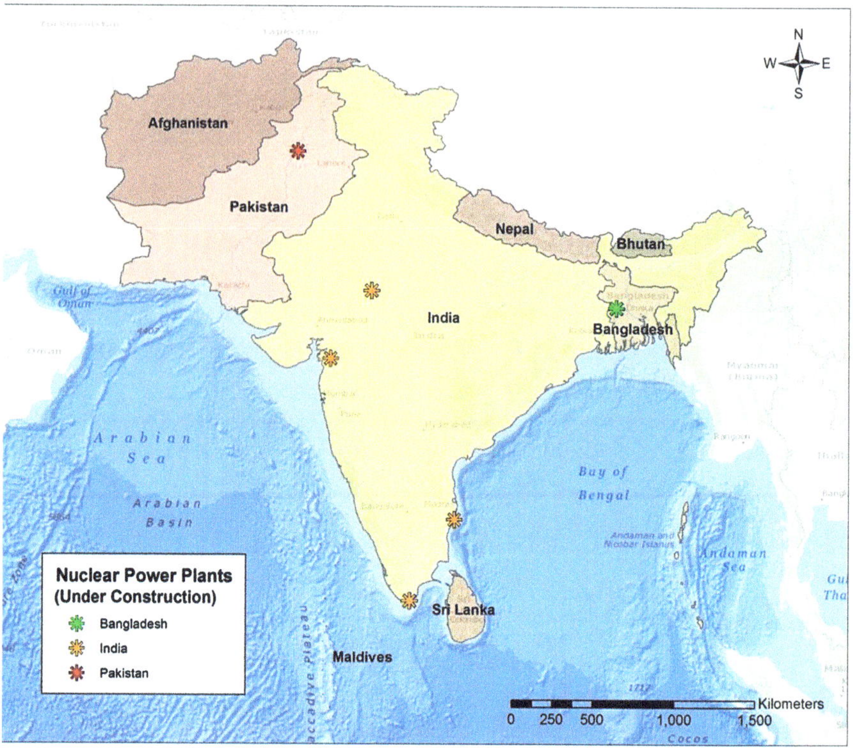

Map 5.4 NPPs under construction. *Source* Results of technical study

and 500 km risk radius was undertaken covering IAEA ingestion risk zones and also based on experiences in Chernobyl and Fukushima.

The result from the extended risk zones, i.e. 300 km shows possibility of ingestion risk from the NPPs in Bangladesh, India and Pakistan. An accident at Rooppur NPP would likely have consequences in India (under 300 and 500 km), and Bhutan and Nepal comes under the 500 km zone. In respect to Kudankulam, as explained in Map 5.3, ingestion risk under the 300 and 500 km category covers Sri Lanka. An accident from the Chashma NPP of Pakistan will have a transboundary ingestion risk (300 km) for Afghanistan and 500 km risk for both Afghanistan and India. Depending on the extent of the ingestion risk under 300 km, it may require evacuation of persons from the affected areas. In other cases of the under-construction plants, the risk appears minimal—varies between remote probability (300 km) to a 'no risk' (500 km).

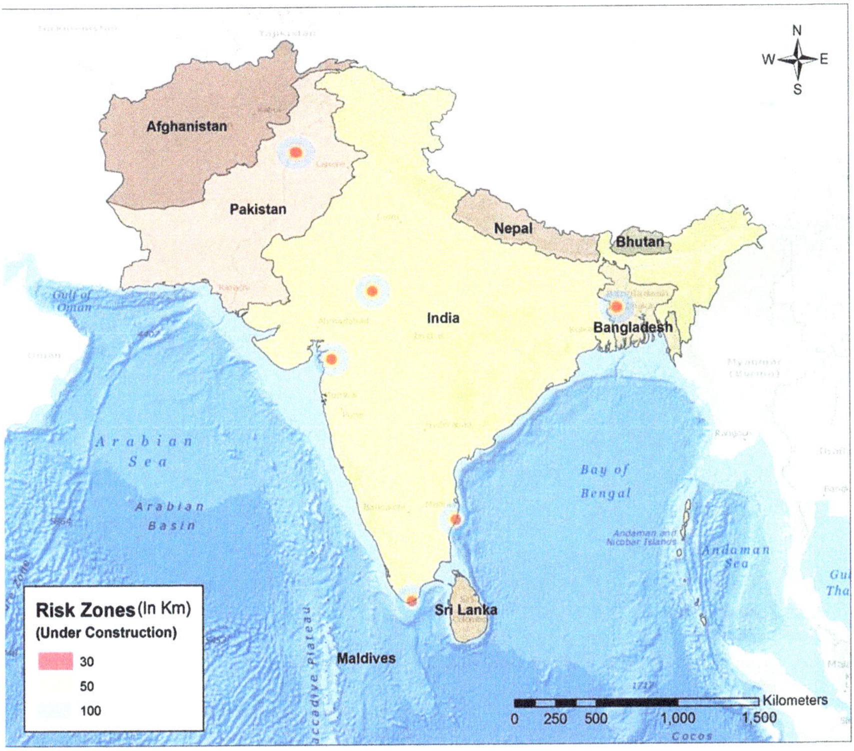

Map 5.5 Risk zones—under-construction NPPs. *Source* Results of technical study

5.3.3 The Proposed NPPs in South Asia

Apart from India and Pakistan, Bangladesh and Sri Lanka—the earlier non-nuclear energy nations, are in the process of setting up of NPPs. Bangladesh has concluded negotiations to setup two plants with a capacity of 1000 MW each at Rooppur, with an option of adding another 2000 MW at the same site in the future. There are reports that suggest Bangladesh is even considering setting up more power plants in the islands of southern region of the country. Sri Lanka has initiated discussions with the IAEA, and as per the news reports (Headlines Today Bureau, 24 July 2012), the site may be located at Sampur in Trincomallee. As the location of the proposed NPP has not yet been officially declared by the government, for the purposes of this study an additional (second) hypothetical location in the extreme south of Sri Lanka has been taken. This second location has been

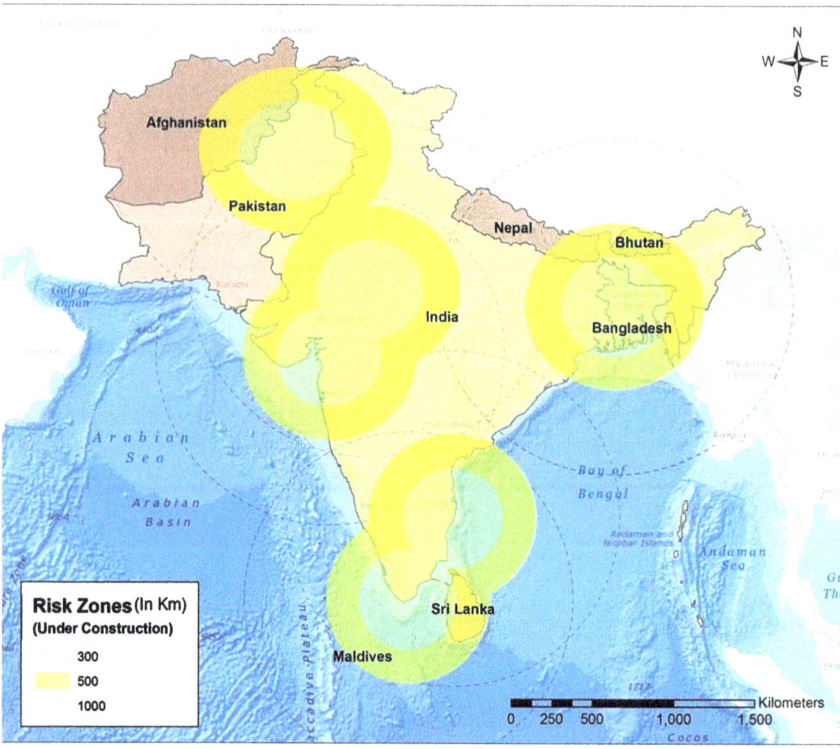

Map 5.6 Extended risk zones plotted for under-construction NPPs. *Source* Results of technical study

chosen to assess what the impact of an NPP, located as far as possible from its neighbouring countries, would likely have.

India's expansion plan is ambitious. It plans to add approximately 30,000 MW by 2020 and 60,000 MW by 2032 from its current capacity of 4800 MW.[13] Pakistan proposes to add two plants of 1000 MW each in Karachi, and several site locations have been identified for the proposed expansion (Press Trust of India 23 April 2012). The site locations are, Taunsa-Punjnad canal near Muzzafargarh; Qadirabad Headworks; Dera Ghazi Khan canal near Taunsa Barrage; Taunsa-Punjnad canal near Multan; Nara canal near Sukkur; Pat Feeder canal near Guddu and Kabul River near Nowshera (Map 5.7).

The risk zones were mapped (Map 5.8) according to the officially declared locations and site proposals, and also hypothetical sites with respect to Sri Lanka.

[13] See Chap. 4.

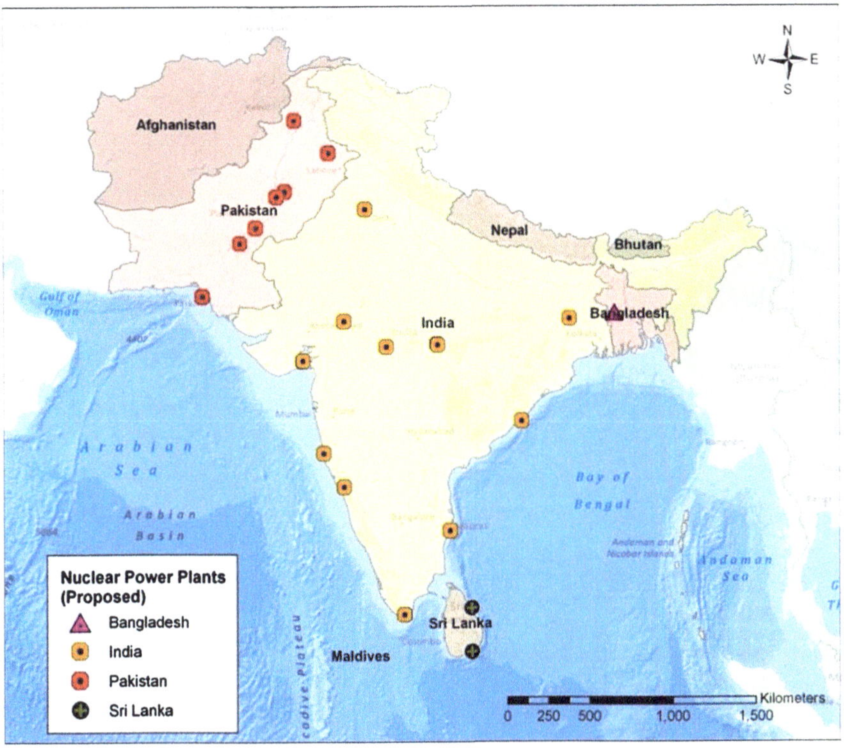

Map 5.7 Planned and proposed NPPs in South Asia. *Source* Results of technical study

The risk mapping of the proposed NPPs in South Asia provides the first concrete evidence of the existence of a risk, both actual and probable.

Rooppur in Bangladesh as explained above in Map 5.5 is close to the international border and an accident will have transboundary consequences in the state of West Bengal, India, which is 50 km aerially from the international border with India. The proposed projects in Pakistan—Qadirabad Headworks; Nara Canal near Sukkur and Pat Feeder Canal near Guddu, in case of accident may likely to have an impact in India under 100 km zone. The Kudankulam NPP will raise similar issues with Sri Lanka, while the proposed NPP in Haripur will lead to transboundary issues with Bangladesh.

Under the conservative gradation of emergency zones, it is obvious from Map 5.8 that the nuclear energy programmes that are being planned in India, Pakistan and Bangladesh would positively require a regional risk approach or reciprocal risk mitigation arrangements among the countries.

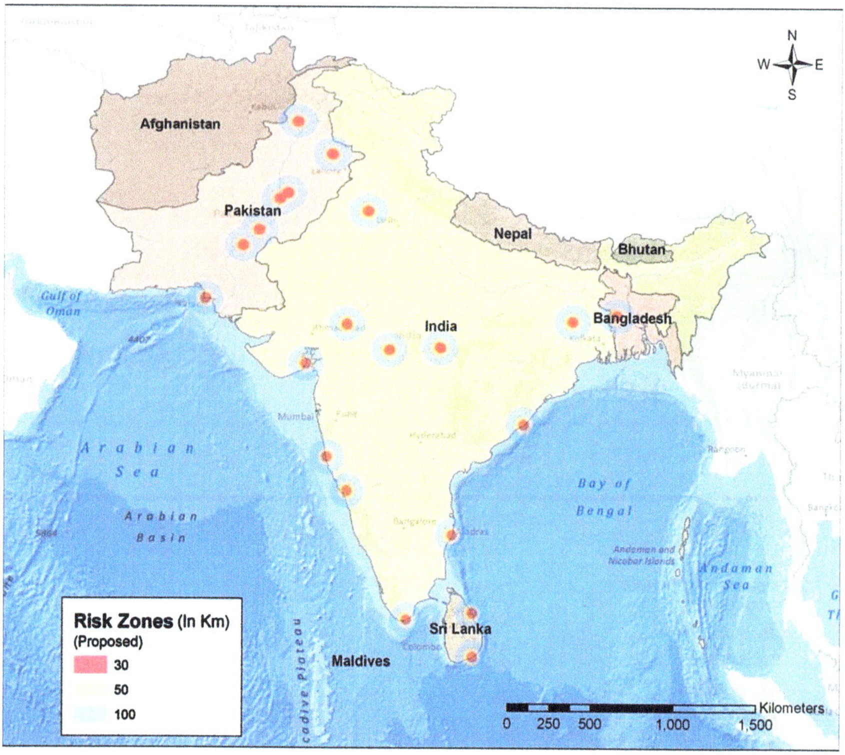

Map 5.8 Risk zones—proposed NPPs. *Source* Results of technical study

Extended risk zone from the proposed NPPs are a matter of regional concern (Map 5.9). Many future projects detailed in Map 5.7 have high-capacity NPPs in the same location and are coming up relatively close to the borders and also to the sea.

Map 5.9 results illustrate that most of the region covered within the extended risk zone, and an accident in any facility depending on the location, will have a direct impact in other parts of South Asia. Maps 5.8 and 5.9 illustrate the importance of regional attention to this issue.

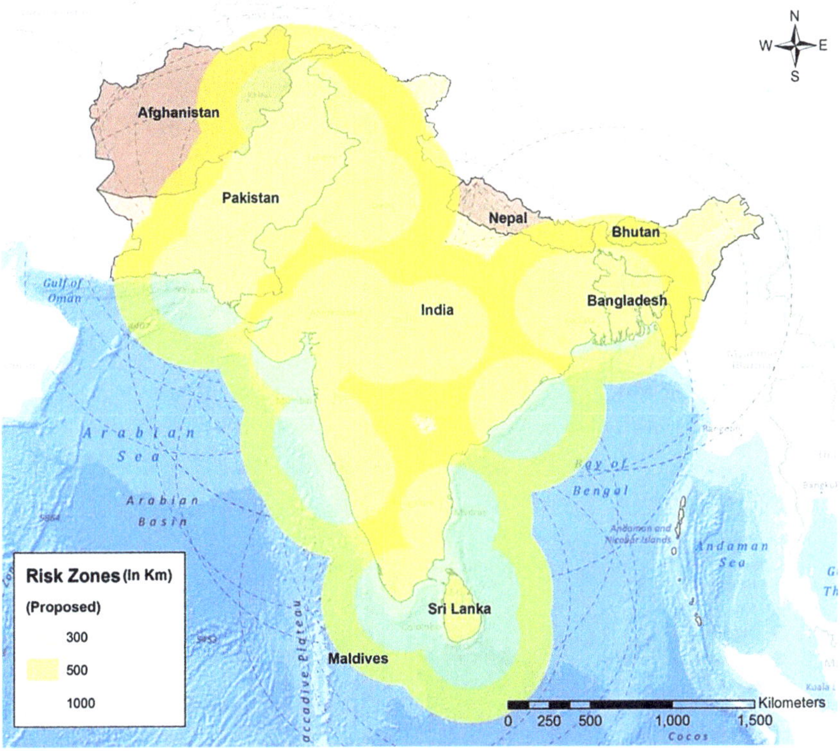

Map 5.9 Extended risk zone from the proposed NPPs. *Source* Results of technical study

5.3.4 Combined Risk Zones (Existing, Under-Construction and Proposed NPPs)

Map 5.10 covers combined location of operational, under-construction and future planned NPPs and Map 5.11 maps the overall impact consisting of all the risk zones from all the operational, under-construction and proposed NPPs.

As Map 5.11 shows, the South Asian region is a nuclear risk zone. All countries of SAARC, except the Maldives, subsequent to an unfortunate accident would be

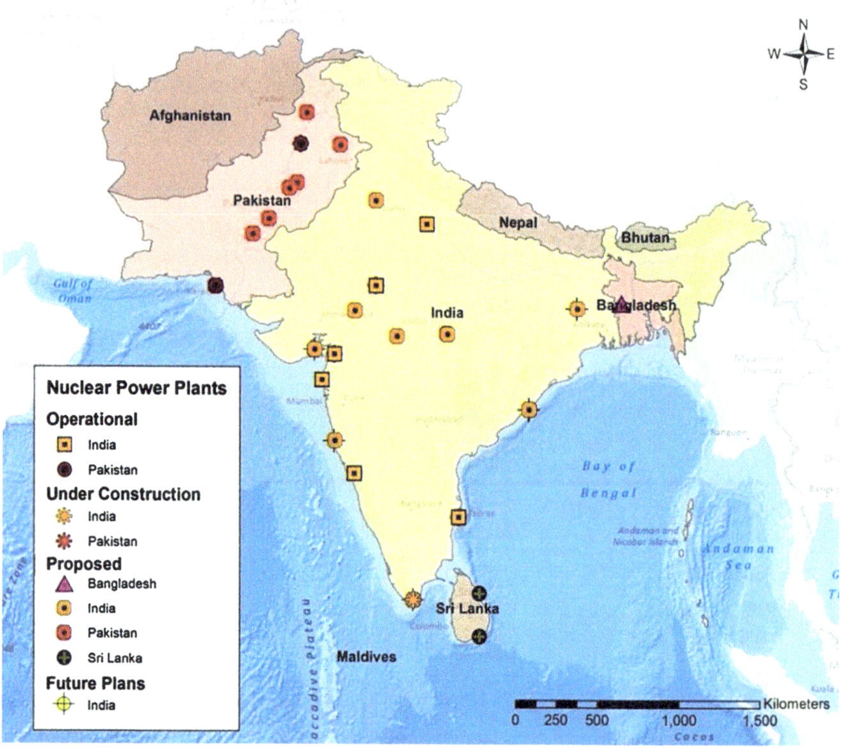

Map 5.10 Location of existing, under-construction, proposed NPPs. *Source* Results of technical study

dealing with shared regional concerns. A transboundary nuclear accident in any of nuclear power countries will have consequences in non-nuclear countries—Afghanistan, Bhutan and Nepal. Many of these countries have no capacity to deal with radiation-related disaster management. If the results of 1000 km radius were to be taken into consideration even the Maldives will come under nuclear emergency zones.

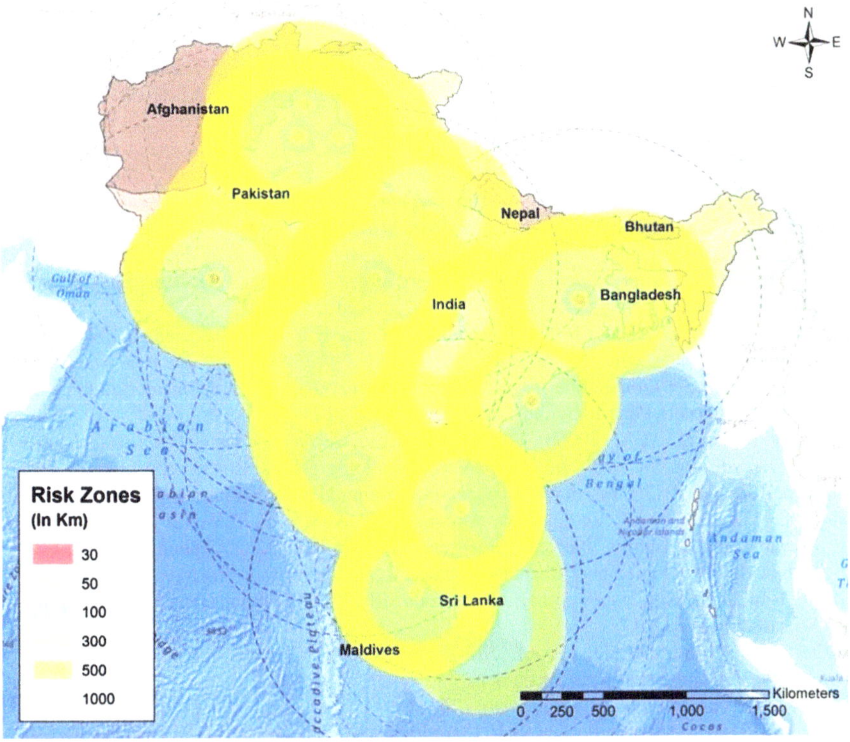

Map 5.11 Combined risk zones (existing, under-construction, proposed). *Source* Results of technical study

5.4 Conclusion

The results of the mapping exercise show in case of accident, depending on the intensity, the South Asian region can be classified as a nuclear risk zone. This means that nuclear power producing countries and also countries that are planning to have NPPs need to consider transboundary risk arising from an accident. The conservative risk (30, 50 and 100 km) results show probability of a transboundary risk between India and Pakistan; India and Sri Lanka; and India and Bangladesh. In the case of ingestion risk (300 km), results demonstrate existence of a transboundary impact in South Asia with the exception of Maldives. A liberal interpretation with 500 and 1000 km would cover the whole SAARC region.

References

Atomic Energy Regulatory Board (AERB). (2000). *Preparedness of the operating organisation for handling emergencies at nuclear power plants*, Safety guide no. AERB/SG/O-6, AERB, Mumbai.

AERB. (2005). *Site considerations of nuclear power plants for off-site emergency preparedness*, Safety guide no. AERB/NPP/SG/S-8, AERB, Mumbai.

AERB. (2009). *Preparedness of the operating organisation for handling emergencies at nuclear power plants*, Safety guide no. AERB/SG/O-6, AERB, Mumbai.

Buck, E.H., & Upton, H.F. (2012). *Effects of Tohoku Tsunami and Fukushima Radiation on the U.S. Marine Environment*, Congressional Research Service, United States. http://www.fas.org/sgp/crs/misc/R41751.pdf. Accessed 2 Aug 2012.

Chernobyl Forum & IAEA. (2006a). *Chernobyl's Legacy: Health, Environmental and Socio-economic Impacts and Recommendations to the Governments of Belarus, the Russian Federation and Ukraine*, International Atomic Energy Agency. www.iaea.org/Publications/Booklets/Chernobyl/chernobyl.pdf. Accessed 4 July 2012.

Chernobyl Forum & IAEA. (2006b). *Environmental consequences of the Chernobyl accident and their remediation: twenty years of experience*, Radiological assessment reports series, IAEA, viewed June 4, 2012, from http://www.pub.iaea.org/mtcd/publications/pdf/pub1239_web.pdf.

Dauer, L.T., Zanzonico, P., Tuttle, R.M., Quinn, D.M., & Strauss, H.W. (2011). The Japanese Tsunami and Resulting Nuclear Emergency at the Fukushima Daiichi Power Facility: Technical, Radiologic, and Response Perspectives, *Journal of Nuclear Medicine*, 52(9), 1423–1432. http://jnm.snmjournals.org/content/early/2011/07/28/jnumed.111.091413.full.pdf. Accessed 15 May 2012.

Fairlie, I. L., & Sumner, D. (2006). *The Other Report on Chernobyl (TORCH)*. Brussels: Greens/EFA in the European Parliament.

Federal Emergency Management Authority. (2002).

Green, H. P. (1973). Nuclear Power: Risk, Liability, and Indemnity. *Michigan Law Review, 71*(3), 479–510.

Greenpeace. (2006). *The Chernobyl Catastrophe Consequences on Human Health*, Greenpeace, Amsterdam.

IAEA. (2002). *Preparedness and Response for a Nuclear or Radiological Emergency: Requirements No. GS-R-2*, IAEA safety standards series, IAEA. http://www-pub.iaea.org/MTCD/publications/PDF/Pub1133_scr.pdf. Accessed 20 April 2011.

IAEA. (2003). *Method for Developing Arrangements for Response to a Nuclear or Radiological Emergency Updating IAEA TECDOC 953*. Emergency preparedness and response series, IAEA. http://www-pub.iaea.org/mtcd/publications/pdf/method2003_web.pdf. Accessed 22 April 2011.

IAEA. (2007). *Arrangements for preparedness for a nuclear or radiological emergency: Safety Guide No. GS-G-2.1*, IAEA safety standards series, IAEA. http://www-pub.iaea.org/mtcd/publications/pdf/pub1265_web.pdf. Accessed 12 June 2012.

Kasperson, R.E., Berk, G., Pijawka, D., et al. (1980). Public opposition to Nuclear Energy: Retrospect and Prospect, *Science, Technology, & Human Values*, 5(31), 11–23.

Kubanyi, J., & Kirchsteiger, C. (2005). *Risk-informed Support of Nuclear Power Plant Emergency Zoning: Benchmarking and Harmonising Strategic Planning Practices for NPP Emergency Zoning and Information to the Public*, DG JRC European Commission. http://publications.jrc.ec.europa.eu/repository/bitstream/111111111/5848/1/EUR%2021580%20EN.pdf. Accessed 4 May 2011.

Office of Nuclear Regulator (ONR). (2015). *ONR principles for determination of offsite emergency planning areas (formerly known as DEPZs)*, viewed March 20, 2015, from http://www.onr.org.uk/depz-onr-principles.htm.

Press Trust of India. (2012). Pakistan plans to build two nuclear power plants in Sindh, *NDTV*, 23 April. http://www.ndtv.com/article/world/pakistan-plans-to-build-two-nuclear-power-plants-in-sindh-201022. Accessed 5 June 2012.

Prime Minister's Office, Japan. (2011). *Press briefing for members of the foreign press*, 22 April. http://www.kantei.go.jp/foreign/incident/pdf/foreign-press-briefing-20110422-script-e.pdf. Accessed 20 May 2012.

Rekacewicz, P. (2007). *Radiation from Chernobyl*, UNEP/GRID-Arendal. http://www.grida.no/graphicslib/detail/radiation-from-chernobyl_7925. Accessed 18 Dec 2011.

Roshan, A.D., Shylamoni, P., & Acharya, S. (n.d). *Monograph on siting of nuclear power plants*, AERB, Mumbai.

Seoul ready to send planes, ships to evacuate its citizens from Japan. (2012). *The Korean Herald*, 18 March. http://nwww.koreaherald.com/view.php?ud=20110318000919. Accessed 12 Dec 2012.

Stellfox, D. (2012). US NRC will review evacuation zone expansion, but separate from Fukushima review, *I-Nuclear.* http://www.i-nuclear.com/2012/05/04/us-nrc-will-review-evacuation-zone-expansion-but-separate-from-fukushima-review/. Accessed 20 Oct 2012.

Sugimoto, A., Krull, S., Nomura, S., et al. (2012). The voice of the most vulnerable: lessons from the nuclear crisis in Fukushima, Japan, *Bulletin of the World Health Organization*, 90(8), 629–630. http://www.who.int/bulletin/volumes/90/8/11-094474/en/index.html. Accessed 20 July 2012.

The International Chernobyl Project. (1991). *Technical report: Assessment of radiological consequences and evaluation of protective measures*, International Atomic Energy Agency. http://www-pub.iaea.org/MTCD/publications/PDF/Pub885e_web.pdf. Accessed 4 Aug 2012.

United Nations Scientific Committee on the Effects of Atomic Radiation (UNSCEAR). (2000). *Exposures and effects of the Chernobyl Accident Annex J*, United Nations. www.unscear.org/docs/reports/annexj.pdf. Accessed 26 June 2012.

UNSCEAR. (2008). *Report to the General Assembly, Annex D.* http://www.unscear.org/docs/reports/2008/11-80076_Report_2008_Annex_D.pdf.

UNSCEAR. (2012). *Interim Findings of Fukushima-Daiichi Assessment presented at the Annual Meeting of UNSCEAR*, United Nations. http://www.unis.unvienna.org/unis/pressrels/2012/unisous144.html. Accessed 11 Oct 2012.

US Nuclear Regulatory Commission. (1993). *Code of Federal Regulations.*

Chapter 6
Establishing a South Asian Nuclear Risk Community: An Empirical Analysis

Abstract The establishment of SAARC is a significant political initiative within the politically diverse, but culturally linked South Asia. SAARC's aim is to achieve South Asian regional integration through mutual trust, understanding and appreciation of each other through economic, technological, social and cultural cooperation, emphasising on collective self-reliance. This being the case, the expansion of the nuclear energy programme in South Asia represents an emerging area of cooperation. With this background, interviews were conducted with diplomatic missions of SAARC countries in Delhi, nuclear policy makers and officials from the SAARC institutions. The purpose of the study is to understand the risk perception of the nuclear energy programme in South Asia. This chapter presents the results of the empirical study and concludes that SAARC nations clearly are willing to explore regional approaches at the SAARC level.

Keywords South Asia and nuclear energy · Shared regional concern on nuclear energy expansion and transboundary risk · Expert assessment of nuclear risk in South Asia · South Asia Association for Regional Corporation and nuclear risk

6.1 Introduction

The literature review and analysis of the SAARC nuclear energy programmes in Chap. 4 brings to light the future expansion of nuclear energy programmes by countries within South Asia. This expansion is already underway without a corresponding transnational legal and institutional framework. The mapping exercise undertaken in Chap. 5 provides a technical background to the nature of nuclear risk within the region.

M.P. Ram Mohan, *Nuclear Energy and Liability in South Asia*,
DOI 10.1007/978-81-322-2343-6_6

With the analysis and results at hand, interviews[1] were conducted to understand how the risks were perceived by policy makers. The purpose of the interviews was to identify their views on a viable institutional system within the SAARC, for the management of risk. A qualitative method of analysis was adopted. The interviewees were experts, ranging from diplomats with prior experience in SAARC issues, policy makers, nuclear scientists, academicians, lawyers, science reporters and disaster management experts.

The subject of nuclear energy is such that an enquiry on the subject, in most cases evokes more passion than reason. The qualitative method is better suited for analysing powerful institutions that make policy, law, and are drivers of change. A key person interview method was employed, and the results were analysed against the framework of the existing literature on nuclear liability, the SAARC structure and technical results.

The interview exercise conducted during 2011–2013 became a crucial component of the research in understanding how the nuclear energy community and experts view the South Asian nuclear energy programmes, especially in the post-Fukushima environment. The interview method was undertaken to gauge the perception of nuclear and institutional experts, towards addressing the consequences of transboundary nuclear accident and its mitigation.

The researcher undertook a visit to the Narora Atomic Power Station (NAPS) in Bulandshahar District, Uttar Pradesh (21 January 2012)[2] for the purpose of understanding the operation of a nuclear plant and also to hear the views of plant officials on safety. The site visit was facilitated by the Nuclear Power Corporation of India Limited (NPCIL) through its public outreach programme. At the plant site, meetings were carried out with technical experts and public communication specialists.

While introducing the research subject, nuclear expansion programmes in the SAARC countries were brought to the notice of the interviewees. The issue of nuclear transboundary risk has not been previously discussed at any SAARC forum. Based on the area of expertise of the interviewees, the interview was conducted through a questionnaire commonly modelled, but with slight difference as per the requirements of specific interviews. Wherever personal interaction was not possible, email interviews were conducted.

There were some limitations and reservations to the interviews conducted. Repeated attempts to interview Pakistani officials in the scientific and diplomatic community did not yield any response. Also, the intent of the research is not to detail any institutional structure, but to bring out the preference for a broad

---

[1] The names of the interviewees are not provided in the book. The interviews were conducted by the Researcher with officials from government, public sector companies, nuclear research Institutes and Embassies. They clearly indicated that views and opinions expressed are for academic discourse only. The view does not represent official positions of any country or organisation.

[2] Narora Atomic Power Station Narora (NAPS) is situated in Bulandshahar District, Uttar Pradesh, about 150 kms from Delhi. NAPS is a unit of the Nuclear Power Corporation of India Limited (NPCIL), with a capacity of 440 KW (2 × 220 MW).

acceptable framework. Moreover, though acceptability of a reciprocal (bilateral) arrangement was also discussed during the course of interview, many felt that India would become a dominant factor where the SAARC will fail to have any role.

6.2 South Asia's Quest for Nuclear Energy

The global nuclear disarmament regime under the NPT facilitates access to nuclear technology to member countries without any discrimination. The countries are within their sovereign right to pursue nuclear energy programmes, as long as they are in compliance with international safeguards. In the South Asian case, India and Pakistan are non-members of NPT, making the region unique.

The interviewees were unanimously of the view that every country in South Asia has a right to determine energy choices, based on their national requirements and priorities. With respect to availability of nuclear energy, countries in SAARC can be classified as follows:

- Countries possessing civilian and military nuclear technology (India and Pakistan);
- Countries in the process of obtaining civilian nuclear technology (Bangladesh and Sri Lanka);
- Countries without nuclear technology (Bhutan, Nepal, Afghanistan and Maldives).

In the pre-Fukushima environment, the South-Asian countries planned a major nuclear energy expansion. Post the Fukushima disaster, referring to India's plan, the senior most official of the Atomic Energy Commission (AEC) (22 February 2012) during a personal communication, on the side-lines of the World Nuclear Energy Conference in New Delhi (22–23 February 2013) stated that "there is no turning back on India's nuclear energy quest; halting India's nuclear energy growth will not be in interest of India's technological pursuit or economic growth". However, he conceded that growth of nuclear energy programmes after the Fukushima disaster is a challenge, and that there is a need to incorporate additional safety features, and better engagement with the public to gain acceptability. The official from the Homi Bhabha National Institute (HBNI) (23 March 2012) also supported this argument. Referring to a study by Grover and Chandra (2006), he stated that without nuclear energy, India's electricity requirements would be difficult to be met.

Considering the fact that except Germany, no major expansion plans were halted, there are many reasons to suggest that nuclear energy will remain a potential energy source (*former AEC official*, 22 February 2012). The Government of India is of the view that India should continue on the path of achieving energy independence, and nuclear energy has a major role to play towards the same. This was reiterated by the new government when the newly elected Prime Minister visited the Bhabha Atomic Research Centre, and in his speech observed the "essential role for nuclear energy in India's energy strategy, given the scale of demand in India" (21 July 2014).

A senior legal officer from the BAEC (18 February and 26 July 2012) informed that the official position of Bangladesh was to have a major share of energy from nuclear in the future. Towards this, he reported that the country had enacted legislation and also cleared construction activities in setting up two NPPs in collaboration with Russia. Further, the Government of Bangladesh was also planning to have more nuclear power plants. This was corroborated by the legal officer; who stated that the government had already constituted an inter-ministerial working group for selection of the sites (28 September 2014).

A high-ranking official from the Sri Lankan High Commission in New Delhi (21 November 2012) suggested that the current demography and economy of Sri Lanka was such that it did not immediately require establishing a nuclear power plant. Nevertheless, by 2025 or 2030 there were all possibilities that Sri Lanka may require an NPP. According to the official, the Government of Sri Lanka is already in consultation with IAEA on this possibility.

In the case of Nepal, the 2007 National Nuclear Policy provides an option to the government to consider nuclear energy in the future. The Political Counsellor at the Embassy of Nepal in India, who is also a former NPT representative of Nepal to the United Nations, was categorical in stating that Nepal has no immediate plans to initiate a nuclear energy programme (02 October 2012). According to him, the country's current economic growth does not demand major changes in power capacity, and there is no political or business rationale to pursuing nuclear energy in the short term. However, he did not rule out the use of nuclear energy in the future.

It is to be remembered that the development of nuclear power in South Asia can be seen as a response to the massive plans for economic development, prolonged power crisis faced by the countries, and the realisation that access to energy is critical for improving lives and developing the economy.

The inability to access the relevant nuclear technology, which could have hindered the domestic policy prescription, does not appear to be a problem anymore. India, Pakistan and Bangladesh, through various channels are dealing with the international community to give shape to their nuclear power generation ambitions. As far as Sri Lanka is concerned; being an NPT member country Sri Lanka can pursue its aspirations, and there is no reason to believe that they would be denied the same. With regard to Bhutan, the official from the Royal Bhutanese Embassy (14 December 2012) was categorical that the country has no intent of initiating a nuclear programme. At the same time, Bhutan would not question the expansion plans of its neighbouring countries. He hoped that any expansion was being undertaken with full comprehension of the risks and responsibilities.

The Science and Environment Correspondent at a prominent business newspaper in India states that as long as a good credible case is put together domestically, even the non-nuclear countries in South Asia can consider nuclear energy as a means of building up their energy capacity, and will soon make known their intentions (24 June 2012). The Counsellor from the Embassy of Nepal acknowledged that, obtaining nuclear energy technology was a long drawn international affair (4 October 2012). Though he was of the view that in the South-Asian context, many countries would like to have a nuclear energy programme; the question being when and how.

6.3 Shared Regional Concerns Within the SAARC

All efforts are being made by Bangladesh, India, Pakistan and Sri Lanka to either expand or initiate nuclear energy programmes. In view of the same, the interviewees were asked about how they viewed the development of nuclear energy in South Asia; if they had any concerns; whether post-Fukushima the countries would be willing to discuss their disaster concerns regionally; and how they would they prefer the concerns to be addressed?

While everyone agreed that it may not be appropriate to comment on the sovereign decision of setting up an NPP by another country, they felt that the governments were obligated to do more to address the safety concerns. Some of the terms commonly used to identify these concerns were, 'risk and safety', 'consequences of an accident' and 'mitigation capabilities'. These answers reflected the genuine or perceived risks associated with the use of nuclear technology, amongst diplomats, scientists and policymakers.

A senior official at the Ministry of External Affairs, India (23 July 2012) conceded that though remote, an unfortunate accident resulting from a nuclear plant in South Asia could be considered as a shared regional concern.

Amongst SAARC countries, Nepal is the only country that has strongly argued the issue of transboundary nuclear energy concerns within South Asia. The National Nuclear policy (2007) of Nepal states "Nepal being surrounded by several nuclear power plants in its neighbour and in the context of their number being continuously increased, Nepal cannot remain untouched from any kinds of nuclear accidents that might occur in its neighbourhood". This policy was strongly supported by a diplomat from the Nepal Embassy (02 October 2012). He explained that the concern was not the expansion of nuclear energy programmes, but the unwillingness of the countries to ask common regional questions as to its safety and overall risk. Since Nepal borders India and might be affected by the fallout of an accident in Bangladesh, a specific question was asked as to the concerns of Nepal. The diplomat from Nepal Embassy had said that such a scenario would necessitate a diplomatic involvement, but as of now nothing is on the table. A senior legal officer from the Bangladesh Atomic Energy Commission (BAEC) opined (18 February 2012) that the expansion of NPPs in South Asia provided a chance for the SAARC community to evaluate the regional risk and act upon it.

The Secretary General of SAARC Law (*an association of the legal communities of the SAARC countries comprising judges, lawyers, academicians, law teachers, public officers,* etc., *with the status of a Regional Apex Body of SAARC*) observed that SAARC as a region needs to ask pertinent questions that affect regional issues. (31 August 2012). An officer from SAARC Documentation Centre (18 July 2012) and the Counsellor from Embassy of Nepal (02 October 2012) similarly agreed that questions should come from within countries, between countries and within region and at international forums. According to them the engagement should be a continuing process and pressure for openness needs to be maintained.

Scientists took a substantially more benign view of the threat arising from functioning of multiple NPPs in the South Asian region. The senior most official from the AEC (22 February 2012) stated, "what is there is the perception of risk and not reality of risk". Though he also observed that the public had every right to contest the safety claims and it was for the government to satisfactorily explain the reasons.

During the course of the researcher's visit to NAPS, scientists at NAPS (21 January 2012), while explaining the working of an NPP, were determined to put across the view that nuclear technology was the safest amongst all other technologies. They argued that, nuclear energy was the cleanest form of energy which did not contribute to any major emissions.

On the question of the probability of an accident, the official in charge of nuclear and radiological emergencies at the NDMA (28 August 2012) observed that though accidents were rare, precaution and continuous monitoring must be up to the norms, and even then there could be instances like Fukushima. He explained, 'human systems' and 'water cooling system' are the most crucial parameters in nuclear plant operation. In the event any of these two fails there is a major problem at hand. The scientist at NAPS (21 January 2012), however, offered an explanation that even if these two probabilities occurred, Indian plants are equipped with additional safety features to handle any such emergencies. Citing the major accident free history of NPCIL plants, they say that the worry is unwarranted.

The scientific responses largely re-emphasised the technological advancement of nuclear energy and the safety features built into the NPPs. On asking about the Fukushima accident, and the possibility of a similar accident occurring elsewhere, the response of the scientists ranged from 'rare incident' to 'this is not a nuclear accident but an unexpected natural disaster leading to nuclear incident', etc. Two high ranking executives from NPCIL (22 February 2012) lent support to the same view. They nevertheless admitted that nuclear energy expansion in South Asia may require transnational dialogue. The official in-charge of HBNI (23 February 2012) stated that, what constitutes a concern is that the expansion should not be at the cost of lax technological and institutional deficiencies.

A question was separately posed to officials from the HBNI and the Ministry of External Affairs about India's transboundary risk, from Rooppur, Bangladesh NPP vis-a-vis Bangladesh concerns related to Haripur, West Bengal NPP. There was, however, no satisfactory response given to the same. On the India–Sri Lanka dispute centring on the KNNP, officials from HBNI and NPCIL informed that they had adequately explained to the Sri Lankan government about the working of the plant, and that there would not be any cross boundary impact either in the course of normal operations or an accident (22 February 2012). The interviewee from the Sri Lankan High Commission (21 November 2012) expressed satisfaction with the explanation received from India. The High Commission official responded that newspapers had wrongly reported that Sri Lanka was not satisfied with India's position. Interestingly, the Sri Lankan official said that India and Sri Lanka were looking for a broader nuclear cooperation in the form of "India–Sri Lanka Comprehensive Civil Nuclear Cooperation". The first discussion had taken place on 12 October 2012 in New Delhi. The second round of talks on the draft text of a comprehensive Agreement on Bilateral Civil Nuclear Cooperation was held during 5–6 May 2014 in Colombo and the final agreement was agreed by both the countries on 16 February 2015 in New Delhi.

On transboundary concerns within South Asia, the interviewees unanimously acknowledged that peoples' concern needed to be discussed regionally, as countries in South Asia expanded their nuclear energy programmes. Interestingly, it has been reported that India and Sri Lanka are in discussion to develop a joint disaster response mechanism on account of KNNP (The New Indian Express, 11 June 2012). This appears to be a confidence building measure, and it is too early to say how much it will help in alleviating Sri Lanka's anxiety. It is likely that India will agree for a KNNP-specific disaster management plan with Sri Lanka. This brings to the forefront the remote possibility of a trade-off between India and Sri Lanka, once Sri Lanka plans its own NPP.

The situation in South Asia is not very different from Europe, and the havoc wrecked by Chernobyl is testament to how non-cooperation between countries aggravates the risk. Today, South resembles a pre-Chernobyl Europe.

On the assessment of risk, a former AEC Chairman (23 February 2012) made a pertinent point, observing "low quantitative risk is not a good enough criteria, and maximum impact in public domain needs to be limited irrespective of the low probability". This view has found support from none other than Shunsuke Kondo, the Chairman of the Japan Atomic Energy Commission. Based on the experience of Fukushima, Shunsuke Kondo during his inaugural address of the IAEA training on 11 July 2012 stated, "prepare ourselves to unexpected, low probability, high consequences event, doubting our common sense in the design and operation of our societies or deliberating situation to be caused by beyond design basis. This lesson gives you freedom to deliberate beyond the expected". Shunsuke's observation summarises the threat that may exist within nuclear establishments, and offers direction in continuously evaluating nuclear risk.

6.4 Disaster Management Initiatives within the SAARC

SAARC has pursued a number of initiatives towards transboundary disaster management. The SAARC Disaster Management Centre (SDC) is based in New Delhi, an independent unit under the SAARC Secretariats at the forefront of this work.

Chapter 4 brought out in detail the disaster management programme at SAARC, and concludes that as of today, all the programmes cover only natural disasters. The official-in-charge of SDC (8 August 2012) stated since there are no proposals at present to include industrial disasters, there have been no discussion on this subject. He also pointed out that the 'SAARC Comprehensive Disaster Management Framework' and 'Regional Agreement on Rapid Response' emphasise regional efforts. According to him, though SDC has no role now on industrial disasters, the agreements referred above are broad in scope and can cover industrial disasters, including nuclear disasters if the member countries wish to do so.

With respect to country-specific nuclear disaster plans, currently India and Pakistan have developed nuclear disaster management plans. This is only partly because no other country in the sub-continent has a functional nuclear programme.

As described above, India and Sri Lanka are looking at the possibilities of a joint disaster management plan.

India's disaster management focal point—the National Disaster Management Authority, has brought out, "National Disaster Management Guidelines—Management of Nuclear and Radiological Emergencies, 2009". Though specific to India, it contains a passing reference, that in case of any nuclear accident emanating either from India or from its neighbouring countries, the concerned country will have to act in accordance with the countries international commitments–referencing IAEA Notification and Assistance Conventions (NDMA 2009). Pakistan's "Regulations on Management of a Nuclear or Radiological Emergency 2008" specifies that "transnational emergency means a nuclear or radiological emergency of actual, potential or perceived radiological significance for more than one State". This Regulation of Pakistan in effect follows the IAEA reference of Precautionary Action Zone (PAZ) or the Urgent Protective action planning Zone (UPZ). A reading of the Regulation of Pakistan conveys, emphasis has been placed on the Emergency Conventions provisions. However, as both India and Pakistan have placed substantive reservations against provisions of these two Conventions, making them redundant to a certain extent (detailed analysis is given in Chap. 4).

As the nuclear programme in the region expands, there is a requirement to build nuclear disaster response capabilities both internally and possibly at the regional level. Currently, worryingly, many countries do not seem to have the capacity or capability to effectively respond to a nuclear disaster. A Counsellor from the Embassy of Nepal was of the opinion (02 October 2012) that a common disaster management system should be seriously considered and put in place. He opined that in the event of a nuclear disaster affecting Nepal, the country would request India for assistance. This could be the same in case for Sri Lanka. Since a nuclear disaster would likely have cross-boundary consequences, SAARC Law official (31 August 2012), and Bangladesh Atomic Energy Commission interviewee (26 July 2012) are of the view that SAARC as a unit could be brought into the picture to respond as a single unit. This would help in developing regional capabilities and ensure, slowly, trust among countries. In order to facilitate a SAARC-led response unit, an SDMC officer observed that in case governments consider bringing nuclear disasters under the ambit of SDMC, the best way to discuss nuclear accident would be within the 'Agreement on Rapid Response' and also take it up within the 'Comprehensive Disaster Management Framework'. This is supported by the Sri Lankan High Commission representative (21 November 2012) that instead of starting afresh, it would be more feasible to bring the same under the existing SAARC disaster management plans.

Putting forward the idea and taking a practical view of the proposal, the officer from MEA, India (23 July 2012) opined that, a discussion on nuclear concerns, involving India and Pakistan required substantial political will, and would depend on the nature of the relationship between the countries in the future. Nevertheless, the officer is positive that shared regional concerns would ultimately weigh in favour of a structured discussion resulting in bringing officials to the negotiating table. In the researcher's view, bringing the governments to discussion itself is a significant achievement.

6.5 SAARC: An Appropriate Forum?

Owing to the volatile nature of regional political relationships, the performance of SAARC can be questioned. A former Secretary General, SAARC (7 December 2012) was candid that in the past SAARC was never considered a serious organisation. In meetings, he observed that there were always major differences in opinion between India and Pakistan, which crippled the effective functioning of SAARC. A Diplomat from the Royal Bhutanese Embassy who was also SAARC desk-in charge at the Embassy (14 December 2012) noted that "SAARC is comparatively a young organisation. India and Pakistan are the backbones in the structure. Mistrust between these two countries have affected SAARC functioning. However, one cannot say SAARC has been ineffective".

On a positive note, currently South-Asian politics seems to be much more cordial. Political engagement within and outside is becoming an important element for achieving peace and development in the region. This optimism is reflected in the last two ministerial meetings (16th SAARC Summit, Thimphu, Bhutan, 2010 and 17th SAARC Summit, Addu City, Maldives, 2011), leading to joint plans of action on global financial crisis, terrorism, and natural and industrial disasters. The 18th summit is being held at Nepal during November 2014 on the theme "Deeper Integration for Peace and Prosperity".

With this background and also since civilian nuclear energy is a subject that has not been discussed at all in any of the SAARC meetings, the interview included a question on the capability of SAARC as an institution to respond effectively to a transnational nuclear accident. Is it viewed as a suitable forum by the political and scientific community?

The HBNI official (23 March 2012) pointed out that presently there was no forum or platform for nuclear energy scientists to discuss issues of mutual interest within SAARC. According to a senior legal officer from BAEC (26 July 2012), Bangladesh is a part of the Forum for Nuclear Cooperation in Asia (FNCA).[3] However, FNCA's role is limited to technical studies with members of NPT, therefore India and Pakistan cannot be part of it. The officer from BAEC strongly believed that SAARC is uniquely positioned to play a major role in nuclear risk management. Expanding on this view, an NDMA official who is responsible for nuclear emergencies (28 August 2012) observed that sharing and discussing ideas of nuclear energy regulatory practices should be encouraged and welcomed at the SAARC level. According to him, since there is a major nuclear energy expansion programme in place, it would be most welcome for the region to have a common

[3] Japan-led FNCA (2012) is a cooperation framework for peaceful use of nuclear technology in Asia. Participating countries are Australia, Bangladesh, China, Indonesia, Japan, Kazakhstan, Korea, Malaysia, Mongolia, The Philippines, Thailand and Viet Nam. Under this framework, view and information exchanges are made on the following fields: (1) Radiation Utilization Development (Industrial Utilization/Environmental Utilization, and Healthcare Utilization), (2) Research Reactor Utilization Development, (3) Nuclear Safety Strengthening, and (4) Nuclear Infrastructure Strengthening.

approach in respect of disaster mitigation as well. Such initiatives would bring confidence to the non-nuclear countries in the region. Non-nuclear energy countries in SAARC in particular need to be informed on the nuclear energy programmes, providing such a forum will remove many doubts from their minds. The NDMA official, however, was of the view that such discussions must be based on the highest standards of technical expertise alone and politics should not intervene.

The SAARC Law official (31 August 2012) stated that the whole process first needs to be a political initiative, only then can mobilisation of public opinion happen, allowing for a strong case to be made at SAARC level.

The researcher believes that both these process should happen in parallel, and with equal commitment. Otherwise, the whole region may in future be deprived of science-based explanation and its reasons. Technical discussions should be used to explain the nature of risk—be it perception or real. In the normal political discourse, any event triggers discussion. In respect of nuclear power programmes, as there have not been any major transnational issues till date, nothing has been discussed. According to the SAARC Law official (31 August 2012), this does not mean that SAARC nations are not sensitive to the problem. Given a chance to understand the issue, the official claimed that every country in SAARC would be supportive to the idea of creating a common regional nuclear energy risk mitigation mechanism.

A Senior Science Correspondent from a prominent newspaper (23 March 2012) identified public pressure as the most important aspect. He argued that in the absence of public pressure, many important issues were left unanswered during bilateral and multilateral dialogue. The other Science and Environment correspondent (24 June 2012) was blunter in his appraisal. Considering the politics of the South-Asian countries, where there has not been a focus on any major issues, such initiatives would only materialise when there occurs an accident and when countries grapple with its consequences.

As regards law making within the SAARC, the SAARC Law official said that (31 August 2012) there are no black listed topics within the SAARC. Any SAARC country could raise an issue for discussion. However, the final Ministerial Meeting decision is based on consensus. Consensus is mandatory for the final passage of an agreement as well.

The HBNI official (23 March 2012) and the senior executive at NPCIL (23 March 2012) share a common view that South Asia being a common cultural and geographical unit, an institutional structure within SAARC is certainly desirable. The BAEC senior legal officer (26 July 2012) and the Counsellor from the Embassy of Nepal (02 October 2012) fully support the proposal to initiate a nuclear energy forum within SAARC, consisting of nuclear energy specialists and others. The representatives from the Maldives Embassy and the Sri Lankan High Commission stated that in the absence of any such proposal they would not comment on the same. However, both the countries view a discussion on nuclear energy as a trust building exercise within the region that needs to be promoted. The NDMA official observed (28 August 2012), for large-scale deployment of nuclear power programme; nuclear energy countries should consider taking neighbours into confidence as well; SAARC seems an appropriate organisation. The official was, however, equally of the view that in case a discussion was initiated, discussion should not be used to prolong the setting up of the power programmes, but should be treated as a trust building exercise. The MEA official (23 July 2012) considered the SAARC Constitution as comprehensive, arguing that any topic that is regional in nature can be brought in for discussion. The Sri Lankan High Commission official (21 November 2012) pointed out that though SAARC may not be efficient, it had been able to make good progress, and one should invest in its process.

Almost every interviewee had a positive attitude, and viewed the SAARC as the appropriate forum to discuss transboundary concerns emanating from a nuclear accident, this being a shared regional concern. However, there were different views on the structure such an initiative takes, which is discussed in the next section.

6.6 Nuclear Risk Community or Regional Framework Liability Convention

In South Asia, India is a major centre of economic power and a donor and, Pakistan a leading power in the region. Even if all the other member countries view SAARC as the best available forum for taking up the case of transnational nuclear disaster, support from India and Pakistan is crucial to conceive and pursue such an agenda. As discussed above, SAARC relies on consensus-based decision-making. The consensus of all the eight member countries is a prerequisite condition for the successful conclusion of any agenda.

The nuclear energy programmes, even if civilian, remain a sensitive issue in South Asia, particularly to India and Pakistan on account of non-membership of NPT. The MEA official argued against India's acquiescence to any binding obligation in this regard, particularly any arrangement that would constrain the capacity of Indian institutions in functioning independently. The senior legal official from the BAEC (26 July 2012) was of the view that when the proposal is being floated, broad acceptability should be the benchmark. This is supported by others from the Sri Lankan and Nepal missions, who state that all the options should be approached with an open mind, but one should take a pragmatic view of the situation.

The MEA official considers as a first step, the establishment of a SAARC nuclear risk community, stating that the SAARC Constitution is a broad vision document and legally there is no bar to creating such a forum. What is been vehemently opposed, especially in the near future, is the establishment of 'framework liability' convention; in South Asia. The officer says a convention should be taken up, only after the learning's of the working of nuclear risk community are placed. To support the argument, the officer views, even Pakistan may not agree to have such a Convention. This is a very pragmatic view, fully appreciating the diplomatic pulls and pressures. Since there are no such discussions currently, for the idea to gain momentum, first steps should be small. The science and environment correspondent (24 June 2012) agrees with the view that one has to see this in the context of practicality. The correspondent's opinion is that though this topic deserves regional attention, when it comes to the nuts and bolts of the framework convention, i.e. liability amount, the insurance mechanisms, etc., we may run into a rough patch that can derail the whole initiative. If pressure is mounted on India, the MEA official says that India may take the route of bilateral arrangements with each of the SAARC nations.

As Harshe (1999) in his article summarises, "South Asia can also be characterised as an Indo-centric region due to India's overwhelmingly superior power in relations to its neighbours". A point considerably underlined by the MEA official.

A strong voice for creating a Framework Convention came from the official from SAARC Law (31 August 2012). He says that if the idea of a 'framework agreement' is not discussed from the beginning itself, there will not be any incentive for the countries within SAARC to take the next step, i.e. establishing a 'SAARC nuclear risk community' and 'framework agreement on SAARC nuclear liability and compensation'. The official in charge of HBNI (23 March 2012) points that without getting into the nuances of the structure, if a proposal is rightly pursued by India, it would bring tangible results in the region. As the Senior Science Correspondent (10 October 2012) observed, the time has already been lost; India could have engaged Sri Lanka from the beginning, which did not happen.

The experience of Europe, as Reyners (18 February 2012) observes, provides a parallel example. Pelzer (2010) emphasises that the future of the nuclear liability mechanism lies with formation of regional risk communities and not a global architecture.

6.7 Conclusion

The interviewees were clear that development of nuclear energy programmes should be seen as a sovereign decision taken in the best interest of the country, and no other country had the right to interfere. At the same time, every interviewee was unanimous that concerns regarding the proliferation of NPPs and the consequent risk of transboundary radiation impact should be seen as a regional issue. The nuclear energy expansion plan without having any corresponding engagement on the level of risk, warrants regional deliberation.

A consensus opinion exists regarding the option of exploring regional approaches at the SAARC level. However, there were divergent views on the acceptability of a 'SAARC nuclear risk community' and/or 'regional framework convention on nuclear liability'. The interviewees were agreeable to the creation of a nuclear risk community at the SAARC level. However, there is no cohesion on the framework liability. India is strongly opposed to the creation of a regional convention. Judging by India's stand, it is likely that Pakistan too would hold the same view. Presently, since there are no discussions at the SAARC level on nuclear energy; when such a discussion is initiated, the process should be given time to mature. The interview results convey that SAARC nations may be willing to take the initial step in conceiving a proposal to establish a 'SAARC Nuclear Risk Community'.

References

Forum for Nuclear Cooperation in Asia (FNCA). (2012). *Structure of the FNCA*. Retrieved October 16, 2012, from http://www.fnca.mext.go.jp/english/about/e_aboutfnca.html.

Grover, R. B., & Chandra, S. (2006). Scenario for growth of electricity in India. *Energy Policy, 34*, 2834–2846.

Harshe, R. (1999). South Asian Regional Co-operation: Problems and prospects. *Economic and Political Weekly, 34*(19), 1100–1105.

National Disaster Management Authority. (2009). *National disaster management guidelines—Management of nuclear and radiological emergencies*. New Delhi: National Disaster Management Authority.

Pelzer, N. (2010). Main features of the revised international regime governing liability-progress and standstill. In *International nuclear law: History, evolution and outlook*. Paris: OECD.

Chapter 7
Conclusion and Suggestions

The international law on liability and compensation resulting from a transboundary nuclear accident, has developed remarkably over the last sixty years. The changes were initiated as the actual impact and consequences of a nuclear accident became apparent, after the Chernobyl disaster. Any observer of these developments would testify that the changes such as enhanced compensation, enlargement of the geographical scope for claims, and expanded definition of damage etc. are fundamental in nature and have strengthened the regime.

While substantial changes have been made to the regime, countries have also questioned the direction of the reforms. A critical review of the provisions of the revised protocols reveals the gaps that threaten the very purpose behind the reforms—harmonisation. The coverage and applicability of the revised Conventions (the 1997 Protocol to Vienna Convention, the 2004 Protocol to Paris Convention and the CSC) shows that the reforms are based on adherence and reciprocity by the States that are party to the conventions. This goes against the very essence of universal applicability, and the objective of harmonising different legal regimes. The countries outside the treaties have raised concerns regarding these new provisions. They argue that adherence based on these new principles, would be financially ruinous, in case of an accident. Even though, during the negotiations of the 1997 Protocol, the countries had argued against restrictions being imposed on the Protocol's applicability, these concerns were discarded as part of the compromise. Further, the direct application of the priority rule in the 1997 Protocol, and the indirect application of the priority rule by the European States through their domestic regimes expose the very nature of the intention of prioritisation of compensation. Another concern, particularly of the non-nuclear countries, is that, no neutral tribunals have been specified, and the compensation claims would be arbitrated within the incident State's judicial ambit.

M.P. Ram Mohan, *Nuclear Energy and Liability in South Asia*,
DOI 10.1007/978-81-322-2343-6_7

Moreover, the 1997 Vienna Protocol and the 2004 Paris Protocol, which made significant changes to the old regime, have still not witnessed adherence by many nuclear energy countries. So much so, that the 1997 Protocol has not come into force despite the fact that the number of ratifications required was as low as five State parties; the 2004 Protocol has only seen Norway and Switzerland ratifying; and after several long years CSC will come into force by 15 April 2015. The ten contracting parties to the 1997 Protocol do not possess substantial nuclear energy programs. In Europe, where the impact of Chernobyl was acutely felt, it has taken close to 10 years to approve the revised 2004 Protocol. However, the approved protocol, even after close to a decade of its existence, has seen adherence by only two countries—Norway (non-nuclear) and Switzerland (nuclear phase out country).

Instead of asking the countries to adhere to conventions with which they are dissatisfied; new reform proposals should be put on the table. The expectations from the IAEA Action Plan were considerable, but the final recommendations for a 'global nuclear liability regime', in effect, only re-emphasises the need for the adherence to the conventions by more States.

In the absence of a comprehensive, all-encompassing international regime, countries are independently initiating reforms. Austria has enacted a law that is fundamentally different from the way the European states regard the founding principles of international nuclear law. India, a developing country, has enacted a domestic law, which in many respects challenges the established principles. Many experts have argued that the Indian law is inconsistent with the international principles and hence defective. Such arguments warrant rebuttal. From the point of Indian lawmakers and public, the Civil Liability law retains the essence of both domestic jurisprudence and also international principles. CSC itself provides such flexibility, as we have seen in Chaps. 2 and 3. It is argued that the 1997 Protocol and the 2004 Protocol have not been effective due to the defective nature of the founding principles themselves. The Indian liability law represents a step forward in modernising the international nuclear liability regime. The law takes into account the political and legal realities and provides the necessary safety net to a country that has experienced Bhopal Gas Tragedy. As the international nuclear liability regime lay dormant; the Fukushima accident in Japan again brought to the forefront, liability issues.

The developments from the European Union, proposing an EU liability framework, reciprocity provisions in Canadian and US law, and the results of probable accident risk assessment within South Asian Association for Regional Cooperation (SAARC) emphasises the strategic role of regional organizations. Through regional organizations, the objective of a harmonious international nuclear liability law can be met. Since South Asia is seeking to expand its nuclear energy footprint, the region under the SAARC framework is ideally poised to take a lead in this regard.

Eight countries constitute the SAARC. Within SAARC, currently India and Pakistan run both civilian and military nuclear energy programs and have major expansion plans for the civilian programs. Bangladesh and Sri Lanka are keen to initiate their own civilian programs. This expansion is being undertaken without adequate public consultation, and without an understanding of the risk posed domestically and regionally. As a common geographical and cultural unit, South Asia will face serious human and environmental consequences in the event of a transboundary nuclear disaster. The international legal regimes which bind the SAARC countries will not be adequate for dealing with the issues of State liability and responsibility in the case of a disaster. Countries within SAARC, currently adhere to different international liability regimes. India has a domestic law, and has signed the Compensation Convention; Bangladesh has enacted its domestic law in tune with the Vienna Convention; Pakistan may opt for the Compensation Convention and Sri Lanka for the Vienna Convention. Such diverse adherences essentially represent the country positions on the nuclear Non-Proliferation Treaty—India and Pakistan being non-members of the NPT. Thus in South Asia today, a Chernobyl type situation exists with respect to the transboundary liability regime; placing the region at risk of consequences similar to the Chernobyl nuclear accident. The SAARC Charter covers inclusion of such emerging regional issues, and provides for the creation of institutions to deal the shared regional concerns, including a nuclear accident.

In order to appreciate the likely impact of a nuclear accident transcending national boundaries, a technical exercise was undertaken to map the risk posed by the nuclear energy installations in South Asia. The technical exercise plotted emergency and extended risk zones and argues a case for a proactive pragmatic legal and policy prescription. Regarding specific risk within a 30, 50 and 100 km radius; the mapping exercise shows that the probability of transboundary risk exist, from projects in India, Bangladesh and Pakistan. In the case of ingestion risk (300 km), the whole of South Asia, except Maldives is covered. Mapping of remote impact based on the Chernobyl radiation flow (500 and 1000 km radius), shows that all of SAARC lies within the risk zone. The results provide a basic geographic framework for decision makers in SAARC, for implementing measures that address the risk concern.

The result may very well be contested by nuclear scientists due to the fact that the results, as per the IAEA emergency evacuation or stay inside classification, do not pose a major evacuation emergency. However, the intent of this exercise is not only to identify emergency evacuation or immediate ingestion areas, but also, importantly, to highlight in cases where radiation spreads to large areas as witnessed in the Chernobyl and Fukushima accidents, putting a densely populated region, with rich biodiversity at risk. The Researcher appreciates that the technologies used in new NPPs have far better safeguard features. However, a remote, once in a life time, event is still considered as a risk, keeping in view the possible catastrophic impacts.

Interviews with experts from South Asia show an appreciable regional concern. Every interviewee was of the opinion that the expansion of nuclear energy within South Asia is fully the prerogative of a country concerned and should not be interfered with. However, the risk emanating from this expansion affects the region as a whole and is a 'shared regional concern'. Persons from both nuclear and non-nuclear countries were uniformly supportive of the view, that the SAARC, as a regional forum, must be utilised to address this shared regional concern. The non-nuclear energy countries in particular are concerned with the impact they may have to face on account of a radiation exposure. This is a clear indication that nuclear energy expansion should not be at the cost of fear of non-nuclear neighbouring countries. It is to be noted that though this study was initiated before the Fukushima nuclear accident, during the course of the research, the Fukushima accident occurred leading to global apprehension and concerns about nuclear energy.

On the specific institutional structure within the SAARC—a liability framework or a regional forum, saw major differences of opinion. The Indian position is not in favour of establishing a major nuclear energy mitigation convention at present. Moreover, a consensus is required to reach an agreement within the SAARC. All the interviewees are of the view that a nuclear risk community should be the first step. It is the researcher's view that the objective of the study is not to bind countries in accepting unreasonable terms, but to make them appreciate that a regional mechanism works like a peer pressure group, keeping an eye on the slippages and compliances; benefiting the larger interest of the region. Hopefully, the initiation of projects in India, Pakistan, Sri Lanka or Bangladesh would result in more structured discussions between the neighbours and also regionally, providing confidence to the South Asian region as a whole.

7.1 Specific Suggestions

1. The Indian and Austrian liability law have attempted to differentiate/modernise their domestic laws from the founding principles of international nuclear law. Subsequent to Fukushima, in Japan there is a growing interest to make suppliers accountable for liability as well. Many more countries could follow this in future. Instead of dismissing the new laws as flawed and are contrary to the founding principles, the international community must engage such countries.
2. The European Commission's initiative for a European nuclear liability law is a welcome step. Other new nuclear regions like the South Asia, the South East Asia and the Gulf should consider regional frameworks. These initiatives will facilitate a global liability regime through regional efforts, which could pave way to an enable harmonisation of laws.
3. The IAEA Action Plan on Nuclear Safety—Nuclear Liability recommendations that were announced after the Fukushima disaster fall short of any major reforms. The IAEA must provide the International Expert Group on Nuclear Liability (INLEX), the terms of reference to undertake a possible review of the founding principles of the nuclear liability law.
4. Under the 'Comprehensive SAARC Disaster Management Framework', transboundary industrial disasters, in particular nuclear disasters, must be included. Further, the SAARC should include industrial disasters that may have transboundary consequences like nuclear disaster within the scope of the 'SAARC Regional Agreement on Rapid Response'.

5. India and Pakistan are expanding the nuclear program, while Bangladesh is initiating the program and Sri Lanka hope to seek nuclear energy programs in future. Framing clear national nuclear and radiological disaster management plans is an urgent necessity that also appreciates the concerns of neighbouring countries. SAARC must initiate plans to establish a nuclear energy risk community within South Asia to deal with the eventualities arising out of unfortunate consequences of a nuclear disaster.
6. Citing India's law as inconsistent, there is substantial pressure from the international community to amend the provisions of the law, particularly in respect to liability amount, right of recourse (Section 17) and tort claims. Even under severe international pressure wisdom of Parliament should not be diluted.
7. Before India ratifies the Convention on Supplementary Compensation, a Parliamentary debate on its merits should be held. Even though Parliamentary ratification is not required as per law in India, the debate will help all the stakeholders to understand the full implications of the CSC ratification.

7.2 Future Scope of Work

1. Drafting a Framework Agreement on a SAARC Nuclear Energy Risk Community.
2. A limited comparative public consultation exercise with the stakeholders in all the nuclear energy countries within South Asia.
3. A full study of the possible impact on the environment and livelihood, and the consequences of a transboundary nuclear disaster in South Asia.
4. A comprehensive work on regional nuclear risk assessment covering detailed weather conditions (air and water flow), seasonal variations etc. pertaining to the existing and new nuclear power plants.
5. Drafting a guidance document on the future of international nuclear liability principles.

Index

M.P. Ram Mohan, *Nuclear Energy and Liability in South Asia*,
DOI 10.1007/978-81-322-2343-6

CPSIA information can be obtained at www.ICGtesting.com
Printed in the USA
LVOW02*2007050515

437319LV00001B/32/P